STUDENT'S SOLUTIONS MANUAL

ROSE SEBASTIANELLI
University of Scranton

LINDA DAWSON
University of Washington

BUSINESS STATISTICS

SECOND EDITION

Norean R. Sharpe
Georgetown University

Richard De Veaux
Williams College

Paul Velleman
Cornell University

Addison-Wesley
is an imprint of

PEARSON

The author and publisher of this book have used their best efforts in preparing this book. These efforts include the development, research, and testing of the theories and programs to determine their effectiveness. The author and publisher make no warranty of any kind, expressed or implied, with regard to these programs or the documentation contained in this book. The author and publisher shall not be liable in any event for incidental or consequential damages in connection with, or arising out of, the furnishing, performance, or use of these programs.

Reproduced by Pearson Addison-Wesley from electronic files supplied by the author.

ISBN-13: 978-0-321-68940-5
ISBN-10: 0-321-68940-2

1 2 3 4 5 6 EB 15 14 13 12 11

www.pearsonhighered.com

Contents

Chapter 2. Data   2-1

Chapter 3. Surveys and Sampling   3-1

Chapter 4. Displaying and Describing Categorical Data   4-1

Chapter 5. Displaying and Describing Quantitative Data   5-1

Chapter 6. Correlation and Linear Regression   6-1

Chapter 7. Randomness and Probability   7-1

Chapter 8. Random Variables and Probability Models   8-1

Chapter 9. The Normal Distribution   9-1

Chapter 10. Sampling Distributions   10-1

Chapter 11. Confidence Intervals for Proportions   11-1

Chapter 12. Confidence Intervals for Means 12-1

Chapter 13. Testing Hypotheses   13-1

Chapter 14. Comparing Two Groups 14-1

Chapter 15. Inference for Counts: Chi-Square Tests   15-1

Chapter 16. Inference for Regression   16-1

Chapter 17. Understanding Residuals   17-1

Chapter 18. Multiple Regression   18-1

Chapter 19. Building Multiple Regression Models   19-1

Chapter 20. Time Series Analysis   20-1

Chapter 21. Design and Analysis of Experiments and Observational Studies   21-1

Chapter 22. Quality Control   22-1

Chapter 23. Nonparametric Methods   23-1

Chapter 24. Decision Making and Risk   24-1

Chapter 2 – Data

SECTION EXERCISES

SECTION 2.1

1. a) Each row represents a different house that was recently sold. It is best described as a case.
    b) Including the house identifier, there are seven variables in each row.

SECTION 2.2

3. a) House _ID is an identifier (special type of categorical); Neighborhood is categorical (nominal); Mail_ZIP is categorical (nominal – ordinal in a sense, but only on a national level); YR_BUILT is quantitative (units – year), but could also be treated as categorical (ordinal); FULL_MARKET_VALUE is quantitative (units – dollars); SFLA is quantitative (units – sq. ft.).
    b) These data are cross-sectional. Each row corresponds to a house that recently sold so at approximately the same fixed point in time.

SECTION 2.3

5. The real estate data of Exercise 1 are not from a designed survey or experiment. Rather, the real estate major's data set was derived from transactional data (on local home sales). The major concern with drawing conclusions from this data set is that we cannot be sure that the sample is representative of the population of interest (e.g., all recent local home sales or even all recent national home sales).

CHAPTER EXERCISES

7. **The news.** Answers will vary.

9. **Sales.** The description of the study has to be broken down into its components in order to understand the study. *Who*– who or what was actually sampled–months at a major U.S. company; *What*–what is being measured–money spent on advertising ($ thousand) and sales ($ million); *When*–monthly from 2004–2006; *Where*–United States (assumed); *Why*–to compare money spent on advertising to sales; *How*–how was the study conducted–not specified; *Variables*–what is the variable being measured–there are 3 variables–the date, the amount of money spent on advertising which is quantitative, and sales which is quantitative; *Source* – data are not from a designed survey or experiment; *Type*–data are time series; *Concerns*–none.

11. **Sales II.** *Who*– who or what was actually sampled–quarterly data from a major U.S. company; *What*–what is being measured–quarterly sales ($ million), unemployment rate (%), inflation rate (%); *When*–quarterly from 2004–2006; *Where*–United States; *Why*–to determine how sales are affected by the unemployment rate and inflation rate; *How*–how was the study conducted–not specified; *Variables*–what is the variable being measured–quarterly sales ($ million), unemployment rate (%), and inflation rate (%) which are quantitative; *Source* – data are not from a designed survey or experiment; *Type* – data are time series; *Concerns*–none.

13. **MBA admissions.** *Who*–MBA applicants; *What*–sex, age, whether or not accepted, whether or not they attended, and the reasons for not attending (if they did not accept); *When*–not specified; *Where*–a school in the northeastern United States; *Why*–the researchers wanted to investigate any patterns in female student acceptance and attendance in the MBA program; *How*–data obtained from the admissions office; *Variables*—sex, whether or not the students accepted, whether or not they attended, and the reasons for not attending if they did not accept (all categorical) and age (years) which is quantitative; *Source* – data are not from a designed survey or experiment; *Type* – data are cross-sectional; *Concerns*–none.

15. **Pharmaceutical firm.** *Who*–experimental participants; *What*–herbal cold remedy or sugar solution, and cold severity; *When*–not specified; *Where*–major pharmaceutical firm; *Why*–scientists were testing the effectiveness of an herbal compound on the severity of the common cold; *How*–scientists conducted a controlled experiment; *Variables*—there are 2 variables. Type of treatment (herbal or sugar solution) is categorical, and severity rating is quantitative; *Source* – data come from an experiment; *Type* – data are cross-sectional; *Concerns*–the severity of a cold might be difficult to quantify (beneficial to add actual observations and measurements, such as body temperature). Also, scientists at a pharmaceutical firm could

have a predisposed opinion about the herbal solution or may feel pressure to report negative findings about the herbal product.

17. **Vineyards.** *Who*–vineyards; *What*–size (acres), number of years in existence, state, varieties of grapes grown, average case price ($), gross sales ($), and percent profit; *When*–not specified; *Where*–assume United States as state is recorded; *Why*–business analysts hope to provide information that would be helpful to grape growers in the United States; *How*–not specified; *Variables*—size of vineyard (acres), number of years in existence, average case price ($), gross sales ($), and percent profit are 5 quantitative variables. State and variety of grapes grown are categorical variables; *Source* – data come from a designed survey; *Type* – data are cross-sectional; *Concerns*–none.

19. **EPA.** *Who*–every model of automobile in the United States; *What*–vehicle manufacturer, vehicle type (car, SUV, etc.), weight (probably pounds), horsepower (units of horsepower), and gas mileage (miles per gallon) for city and highway driving; *When*–the information is currently collected; *Where*–United States; *Why*–the EPA uses the information to track fuel economy of vehicles; *How*– among the data EPA analysts collect from the automobile manufacturers are the name of the manufacturer (Ford, Toyota, etc.), vehicle type….”; *Variables*— there are 6 variables. Vehicle manufacturer and vehicle type (car, SUV, etc.) are categorical variables. Weight (probably pounds), horsepower (units of horsepower), and gas mileage (miles per gallon) for both city and highway driving are quantitative variables; *Source* – data are not from a designed survey or experiment; *Type* – data are cross-sectional; *Concerns*–none.

21. **Lotto.** *Who*–states in the United States; *What*–state name, whether or not the state sponsors a lottery, the number of numbers in the lottery, the number of matches required to win, and the probability of holding a winning ticket; *When*–1998; *Where*–United States; *Why*–not specified but likely that the study was performed in order to compare the chances of winning the lottery in each state; *How*–not specified but data could be gathered from a number of different sources, such as the state lottery; *Variables*— there are 5 variables. State name, whether or not the state sponsors a lottery are categorical variables. The number of numbers in the lottery, the number of matches required to win, and the probability of holding a winning tickets are quantitative variables; *Source* – data are not from a designed survey or experiment; *Type* – data are cross-sectional; *Concerns*–none.

23. **Stock market.** *Who*–students in an MBA statistics class; *What*–total personal investment in stock market ($), number of different stocks held, total invested in mutual funds ($), and the name of each mutual fund; *When*–not specified; *Where*–a business school in the northeast US; *Why*–the information was collected for use in classroom illustrations; *How*–an online survey was conducted, participation was probably required for all members of the class; *Variables*— there are 4 variables. Total personal investment in stock market ($), number of different stocks held, total invested in mutual funds ($) are quantitative variables. The name of each mutual fund is a categorical variable; *Source* – data come from a designed survey; *Type* – data are cross-sectional; *Concerns*–none.

25. **Indy 2009.** *Who*–Indy 500 races; *What*–year, winner, car model, time (hrs), speed (mph), and car number; *When*–1911-2009; *Where*–Indianapolis, Indiana; *Why*–examine trends in Indy 500 race winners; *How*–official statistics kept for each race every year; *Variables*— there are 6 variables. Winner, car model, and car number are categorical variables. Year, time (hrs) and speed (mph) are quantitative variables; *Source* – data are not from a designed survey or experiment; *Type* – data are time series; *Concerns*–none.

27. **Mortgages.** Each row represents each individual mortgage loan. Headings of the columns would be: borrower name, mortgage amount.

29. **Company performance.** Each row represents a week. Headings of the columns would be: week number of the year (to identify each row), sales prediction ($), sales ($), and difference between predicted sales and realized sales ($).

31. **Car sales.** Cross-sectional are data taken from situations that vary over time but measured at a single time instant. This problem focuses on data for September only which is a single time period. Therefore, the data are cross-sectional.

33. **Cross sections.** Time series data are measured over time. Usually the time intervals are equally-spaced (e.g. every week, every quarter, or every year). This problem focuses on the average diameter of trees brought to a saw mill in each week of a year; therefore, the data are measured over a period of time and are time series data.

Chapter 3 – Surveys and Sampling

SECTION EXERCISES

SECTION 3.1
1.
 a) False. Sampling error cannot be avoided, even when the sample is unbiased. Sampling error is always present when a sample statistic is used to estimate a population parameter.
 b) True.
 c) True.
 d) False. Randomization will match the characteristics in a way that is unbiased. We can't possibly think of all the characteristics that might be important or match our sample to the population on all of them.

SECTION 3.2
3.
 a) Population–Professional food preparers in the U.S.
 b) Sampling Frame–*Chef's Collaborative* membership listing.
 c) Parameter–Proportion of professional food preparers in the U.S. who believe that food safety has improved.
 d) Sampling method–Simple random sample (SRS).

SECTION 3.3
5.
 a) No. It would be nearly impossible to get exactly 500 males and 500 females by selecting people from India at random.
 b) Stratified sampling. This allows GfK Roper to divide the population of India into two groups based on gender (called strata) and then randomly select 500 individuals from each group. The resulting sample will consist of an equal number of males and females.

7. The consumer advocacy group used systematic sampling. To ensure the sample is random, the consumer advocacy group should have started the systematic sampling by selecting the first member from the *Chef's Collaborative* list at random.

9.
 a) Population—Human resources directors of Fortune 500 companies.
 b) Parameter—Proportion who don't feel surveys intruded on their workday.
 c) Sampling Frame—List of HR directors at Fortune 500 companies.
 d) Sample—23% of HR directors don't feel that surveys intruded on their workday.
 e) Method—Questionnaire mailed to all (nonrandom).
 f) Bias—Nonresponse bias. Those that respond are obviously receptive to participating in surveys. Since who responds is related to the question itself, there is likely a difference between those who respond and those who don't respond to the survey with regard to the question.

SECTION 3.4
11.
 a) The population of interest is professional food preparers in the U.S.
 b) Members of the *Chef's Collaborative* who attended the recent symposium on "Food Safety in the 21st Century" that was held in Las Vegas.
 c) The sampling frame is not necessarily representative of the entire group of food preparers. Those who attended the symposium may have different opinions from those who didn't. It is likely that

those who attended the symposium have a special interest in the topic. Moreover, rather than selecting members to call in a random fashion, he started from the top of the list. The list was generated as members enrolled for the symposium; therefore those at the top of the list enrolled early (and perhaps were more enthusiastic about the issue). Consequently, there may be differences in opinion between members near the top of the list and those near the bottom. Finally, the script is biased (leading respondents to answer in a particular way) and may lead to an inflated estimate of the true population proportion who think food safety has improved.

13.

    a) *Question 1* seems appropriately worded, although using the phrase "state-of-the-art" may give the impression that a higher price could be charged for the service. By noting that the monthly cost for the service is less than what it would cost to have a daily cup of cappuccino, *Question 2* predisposes the respondent to agree that $50 is a reasonable price. *Question 2* does not seem to be appropriately worded.

    b) *Question 1* is more neutrally worded than *Question 2*. In addition, *Question 1* addresses the issue more directly: the willingness of customers to pay $50 per month for the new service.

15.

    a) True.

    b) True

    c) False. Measurement error refers to inaccurate responses. Sampling error refers to sample-to-sample variability and is always present when using a sample statistic to estimate a population parameter.

    d) True

17.

    a) This is a multistage design, with a cluster sample in the first stage and a simple random sample in the second stage. Churches are treated as clusters of the population, and three churches (clusters) are selected. Then a random sample of members is selected from each church.

    b) Each church (cluster) is assumed to be representative of the population. This may not be the case if they are different (e.g., in terms of prominent ethnicity of its members such as Italian, Irish, etc.). If each church is not representative of the population, then bias will be introduced at the cluster stage.

CHAPTER EXERCISES

19. Software licenses.

    a) This sample was a voluntary response, not a random sample.

    b) There is no confidence in the estimate sampled. Voluntary response samples are almost always biased, and so conclusions drawn from them are almost always wrong.

21. Gallup.

    a) The population of interest is all adults in the United States aged 18 and older.

    b) The sampling frame, a list of individuals from which the sample will be drawn, consists of U.S. adults with landline telephones, which are the only numbers available for a study like this.

    c) An increasing number within the population (e.g., many college students and others with mobile only service) don't have landline phones, which could create a bias.

23. Alternative medicine.
   a)   Population–All Consumer Union subscribers.

   b)   Parameter–Proportion of Consumer Union subscribers who have used and benefited from alternative medicine.

   c)   Sampling Frame–All Consumer Union subscribers.

   d)   Sample–Subscribers who responded.

   e)   Sampling method–Questionnaire to all subscribers.

   f)   Bias–Nonresponse. Those who respond could have strong feelings about the topic and affect the results.

25. At the bar.
   a)   Population–Adult bar patrons.

   b)   Parameter–Proportion of sample who thought drinking and driving was a serious problem.

   c)   Sampling Frame–All chosen bar patrons.

   d)   Sample–Every 10th person leaving the bar.

   e)   Sampling method–Systematic sampling (every 10th person).

   f)   Bias–Probably biased toward thinking drinking and driving is not a serious problem. The sample consisted of bar patrons leaving the bar. A large percentage of them had something to drink, most likely leading to a biased viewpoint. In addition, bar patrons don't reflect what all adults think about drinking and driving.

27. Toxic waste.
   a)   Population–Soil located near former waste dumps.

   b)   Parameter–Concentrations of toxic chemicals.

   c)   Sampling Frame–Any accessible soil surrounding a former waste dump.

   d)   Sample–Soil samples taken from 16 locations near a former waste dump.
   e)   Sampling method–Not specified how the sample locations were chosen.

   f)   Bias–Not specified how soil sample locations were chosen and therefore cannot assume they were chosen randomly, perhaps accessibility or some other factors. Unless this is known, it is possible that bias can affect the results if soil taken is more or less polluted than a random selection would produce.

29. Quality control.
   a)   Population–Snack food packages.

   b)   Parameter–Proportion of snack food packages passing inspection, weight of bags.

   c)   Sampling Frame–All snack food packages produced in a day.

   d)   Sample–Packages in 10 randomly selected cases, 1 bag from each case for inspection.

    e) Sampling method–Multistage sampling due to a combination of methods. The selection of the 10 cases is a cluster and the sampling selection of an individual bag from each case is probably a random sample, although this is not specified.

    f) Bias–Should be unbiased as long as the individual bag chosen is random. There could be differences in the first bag of a case versus the last bag.

31. Instant poll. The station's faulty prediction is most likely the result of bias. Only people watching the local TV station news have the opportunity to respond. The responders who volunteered to participate may have different viewpoints than those of other voters, who either chose not to respond or didn't have the opportunity to participate (didn't see the news program).

33. Cable company market research.

    a) Sampling strategy is volunteer response. Bias is introduced because only those individuals who see the ad and feel strongly about the issue will respond. The opinions may not be representative of the rest of the public.

    b) Sampling strategy is a cluster of one town selected to be sampled. Bias is introduced because one town may not be representative of all towns.

    c) Sampling strategy is an attempted census, accessing all customers. Bias is introduced because of nonresponse to the mailing survey.

    d) Sampling strategy is stratified by town, selecting 20 customers at random from each town to be surveyed, including follow up. This strategy should be unbiased and representative of the public opinion about the cable issue.

35. Amusement park riders.

    a) This is a systematic sample (every 10th person in line).

    b) It is likely to be representative of all of those waiting in line to go on the roller coaster. It would be useful to compare those who have waited and are now at the front with those who are in the back of the line. Otherwise, survey every 10th person about to board the roller coaster for a more consistent response.

    c) The sampling frame consists of persons willing to wait in line for the roller coaster on a particular day within a given time frame.

37. Another ride. Biases exist because it could be that only those who think it is worth waiting for the roller coaster ride are likely to still be in line. Those who don't like roller coasters or don't want to stay in lines are not part of the sampling frame. Therefore, the poll won't get a fair picture of whether park patrons overall would favor more roller coasters.

39. (Possibly) Biased questions.

    a) This statement is biased because it leads the responder toward yes because of the word "pollute". The word "pollute" conjures up a negative image leading the responder to agree that companies should pay for this behavior. Another way to phrase it would be "Should companies be responsible for costs of environmental cleanup?"

    b) This statement is biased because it leads the responder to no because of the words "enforce" and "strict" that conjure up images that could lead a responder to having negative reaction. Another way to phrase it would be "Should companies have dress codes?"

41. Phone surveys.

    a) It would be difficult to achieve a random sample in this case because not everyone in the sampling frame has an equal chance of being chosen. People with unlisted phone numbers, without phones, and those at work or away from the home at the designated calling time cannot be contacted.

    b) Another strategy would be to generate random numbers and call at random times or select random numbers from the phonebook and call at random times (this doesn't solve the unlisted phone number issue).

    c) In the original plan, families that have one person at home are more likely to be included in the study. Using the second plan, more people are potentially included although people without phones or those not home when called are still not included.

    d) This change does improve the chance of selected households being included in the study.

    e) The random digit dialing does address all existing phone numbers, including unlisted numbers. However, there is still the issue of residents not being home at the time of the call. In addition, people without phones are still left out of the study.

43. Change.

    a) Answers will vary

    b) The parameter being estimated is the true mean amount of change that you carry daily just before lunch.

    c) Population is now the amount of change carried by your friends. The average parameter estimates the mean of these amounts.

    d) The 10 measurements in c) are more likely to be representative of your class (peer group with similar needs) but unlikely for larger groups outside of your circle of friends.

45. Accounting.

    a) Assign numbers 001 to 120 (3 digits required because the maximum number is 120) representing each order in a day. Use random numbers to select 10 transactions to check for accuracy.

    b) Separate the transactions and sample each type (wholesale and retail) proportionately. This would be a stratified random sample.

47. Quality control.

    a) Randomly select 3 cases and then randomly select one jar from each case.

    b) Assign numbers 01 to 20 to cases 07N61 to 07N80 respectively. Then generate three random numbers between 01 and 20 and select the appropriate case. Then assign random numbers 01 to 12 to each of the 12 jars within each case. For each case selected, generate a random number between 01 and 12 and select the corresponding jar within each case.

    c) The method described involves two separate sampling methods and, therefore, it is multistage sampling.

49. Sampling methods.

    a) Yellow pages may not include all doctor listings. If regular line listings are used, the list may include all doctors. If ads are used, not all doctors would be included and the ones with ad would not be typical of all doctors.

    b) This sampling method is not appropriate. The cluster sample chosen (the randomly selected page) will only contain a handful of businesses and maybe only one or two business types.

Chapter 4 – Displaying and Describing Categorical Data

SECTION 4.1

1.

    a)   Frequency table:

None	AA	BA	MA	PhD
164	42	225	52	29

    b)   Relative frequency table (divide each number by 512 and multiply by 100):

None	AA	BA	MA	PhD
32.03%	8.20%	43.95%	10.16%	5.66%

SECTION 4.2

3.

    a)

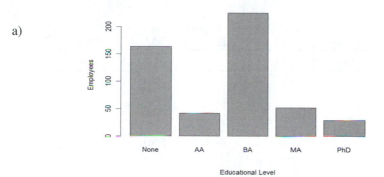

    b)

    c)

5.

    a) Most employees have either a bachelor's degree (44%) or no college degree (32%). About 10% have master's degrees, 8% have associate's degrees, and nearly 6% have PhDs.

    b) It is difficult to generalize these results to any other division of the company or to any other company. These data were collected from only one division. Other divisions and companies might have vastly different educational requirements for their employees and therefore distributions of educational levels.

SECTION 4.3

7.

    a)

	Totals
< 1 year	95
1-5 years	205
more than 5 years	212

    b) Yes.

None	AA	BA	MA	PhD
164	42	225	52	29

9.

    a)

(%)	None	AA	BA	MA	PhD
< 1 year	6.1	7.1	22.2	38.5	41.4
1-5 years	25.6	21.4	49.8	51.9	51.7
more than 5 years	68.3	71.4	28	9.6	**6.9**

    b) No. The distributions look quite different. More than 2/3 of those with no college degree have been with the company longer than 5 years, but almost none of the PhDs (less than 7%) have been there that long. It appears that within the last few years the company has hired better educated employees.

    c)

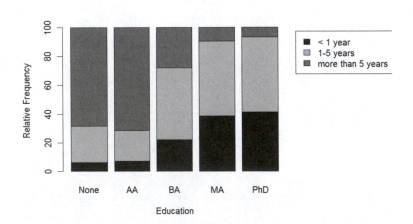

d)   It is easier to see the differences in the distributions in the stacked bar chart.

CHAPTER EXERCISES

11. Graphs in the news. Answers will vary.

13. Tables in the news. Answers will vary.

15. U.S. market share.
   a)   Yes, this is an appropriate display for these data because all categories of one variable (sellers of carbonated drinks) are displayed. The categories divide the whole and the category Other combines the smaller shares.

   b)   The company with the largest share is Coca-Cola.

17. Market share again.
   a)   The pie chart does a better job of comparing portions of the whole.

   b)   The "Other" category is missing and without it, the results could be misleading.

19. Insurance company.
   a)   Yes, it is reasonable to conclude that deaths due to heart OR respiratory diseases is equal to 30.3% plus 7.9%, which equals 38.2%. The percentages can be added because the categories do not overlap. There can only be one cause of death.

   b)   The percentages listed in the table only add up to 73.7%. Therefore, other causes must account for 26.3% of U.S. deaths.

   c)   An appropriate display could either be a bar graph or a pie graph, using an "Other" category for the remaining 26.3% causes of death.

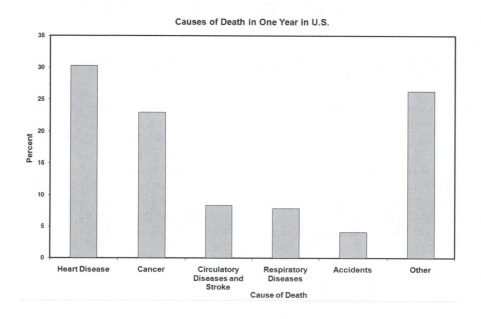

21. Web conferencing. WebEx Communications, Inc. has the majority of the market share for web conferencing (58.4%), and Microsoft has approximately a quarter of the market share. Other companies comprise about 15% of market share so there appears to be opportunity for either company to grow. Either a bar chart or pie chart would be an appropriate graphic display.

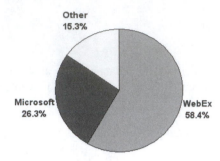

23. Small business productivity.
- **a)** The percentages total more than 100%. The percentages do not represent parts of one variable. They simply represent the percentages of businesses who took certain actions. They could also take more than one action.

- **b)** Bar chart:

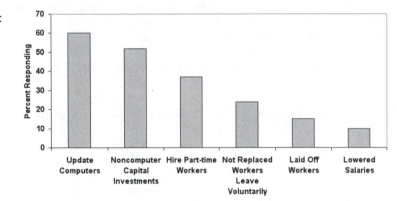

- **c)** A pie chart would not be appropriate because the percentages do not represent parts of a whole and do not total 100%.

- **d)** (Answers will vary) Over 50% of the business owners said that they have updated their computers or made noncomputer capital investments to increase productivity (or both). A smaller percentages of business owners (from 10-37%) made changes in their hiring or salary structures.

25. Environmental hazard. The bar chart shows that grounding is the most frequent cause of oil spillage (118) for these 312 spills, while collisions (97) are ranked a close second. The other causes that with lower values (24-43) were due to hull failures, fires and explosions, and unknown causes. In order to differentiate between close counts, a bar chart is easier to read unless the pie chart gives the actual percentages. It is difficult to determine differences between similar areas in the pie chart. To showcase the causes of oil spills as a fraction of all 312 spills, the pie chart is the better choice.

27. Importance of wealth.
- **a)** India 76.1% - USA 45.3% = 30.8%

- **b)** The vertical axis on the display starts at 40% which makes the comparison between countries difficult and the areas disproportionate. For example, the India bar looks about 5-6 times as big as the USA bar but the actual values are not even twice as big.

c) The display would be improved by starting the vertical axis at 0%, not 40%.

d)

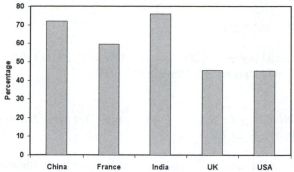

e) The percentage of people who say that wealth is important to them is highest in China and India (over 70%), followed by France (close to 60%) and then the USA and U.K. where the percentages were close to 45%.

29. Google financials.

a) These are column percentages because the column sums add up to 100% and the row percentages add up to more than 100%.

b) A stacked bar chart is appropriate.

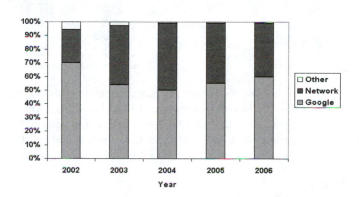

c) The main source of revenue for Google is from their own websites, which in 2002 was 70%, dropping down to 50% in 2004, and back up to 60% in 2006. The second largest source of revenue is from other network websites. While the Google websites have remained the main source of revenue, the revenue from the Google network websites has been increasing. Licensing and other revenue was 6% in 2002 but since 2004 has only been 1%.

31. Stock performance.

a) 62.5% (25/40)

b) 35% (14/40)

c) 15% (6/40)

d) 50% (20/40)

e) 61% (14/23)

f) 65% (11/17)

g) There does not appear to be much, if any, relationship between the performance of a stock on a single day and its performance over the previous year.

33. Real estate.

    a) 10.2% (266/2610); 13.2% (341/2575)

    b) 1.8% (48/2610); 1.5% (38/2575)

    c) The percent of residential real estate sales decreased from 81.2% (2119/2610) in 2006 to 77.9% (2006/2575) in 2007; this represents a decrease of 3.3%.

35. Movie ratings.

    a) Conditional distribution (in percentages) of movie ratings for action/adventure films: G 11.4%; PG 14.3%; PG-13 48.6%; R 25.7%.

    b) Conditional distribution (in percentages) of movie ratings for thriller/horror films: G 0%; PG 0%; PG-13 57.9%; R 42.1%.

    c) Stacked bar chart:

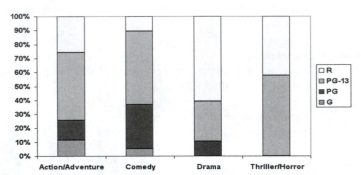

    d) *Genre* and *Rating* are not independent, in other words, *Genre* and *Rating* are related to each other. Thriller/Horror movies are all PG-13 or R and Drama is similar. For example, there is a 5% chance (6/120) that a randomly selected movie is rated G. However, if you were told that the movie was a Thriller/Horror film, there would be a 0% change that it was rated G. Thus, knowing the genre, does affect the rating – they are not independent.

37. MBAs.

    a) 62.7% (168/268)

    b) 62.8% (103/164)

    c) 62.5% (65/104)

    d) The marginal distribution of origin: 23.9% from Asia; 1.9% from Europe; 7.8% from Latin America; 3.7% from the Middle East; 62.7% from North America.

    e) The column percentages:

	Two-Yr	Evening	Total
Asia/Pacific Rim	18.90	31.73	23.88
Europe	3.05	0.00	1.87
Latin America	12.20	0.96	7.84
Middle East/Africa	3.05	4.81	3.73
North America	62.80	62.50	62.69
Total	100.00	100.00	100.00

    f) They are not independent. For example, there is less than a 19% chance (31/164) that a randomly selected Two-Year MBA student is an Asian/Pacific Rim student . However, there is more than a 31% chance (33/104) that a randomly selected Evening MBA student is an Asian/Pacific Rim student. This is over a 50% increase in the likelihood that a student is an Asian/Pacific Rim student. Thus knowing the kind of MBA program does affect the likelihood of the origin of the MBA student.

39. Top producing movies.

    a) 6.0% (12/200)

    b) 5.0% (1/20)

    c) 3.5% (7/200)

    d) 56.7% (34/60)

    e) 73.8% (59/80)

    f) PG and PG-13 films increased from 72% in 1999-2003 to 80% in 2004-2008 while R-rated films decreased from 22% in 1999-2003 to 14% in 2004-2008.

	G	PG	PG-13	R	Total
2004-2008	6.0%	23%	57%	14%	100.0%
1999-2003	6.0%	19%	53%	22%	100.0%

41. Tattoos. The study by the University of Texas Southwestern Medical Center provides evidence of an association between having a tattoo and contracting hepatitis C. Approximately 33% of the subjects who were tattooed in a commercial parlor had hepatitis C, compared with 13% of those tattooed elsewhere, and only 3.5% of those with no tattoo. If having a tattoo and having hepatitis C were independent, we would have expected these percentages to be roughly the same.

43. Revenue growth, last one.

    a) 8%

    b) No, because we aren't given counts or totals. The percentages are within revenue categories.

    c) 92% (48% + 44%)

    d) Although the percent with only a high school education is "relatively" constant over the three revenue categories, the study shows that firms having higher revenues had a significantly higher percentage of women CEO's with graduate educations.

45. Moviegoers and ethnicity.

    a) Hispanic 14.5%; African-American 12.5%; Caucasian 73.0%.

    b) For 2006, Hispanic 15.7%; African American 12.0%; Caucasian 72.3%.

c)

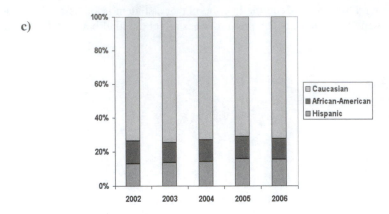

d) The conditional distribution of *Ethnicity* is almost the same across the five *Years*; however, there seems to be a slight increase in the percentage of Hispanics who go to the movies from 13.1% in 2002 to 15.7% in 2006.

47. Women's business centers.

   a) Row percentages (add up to 100%).

   b) The stacked bar chart compares the locations of women's business centers. The result is that the percentages are similar for both *Established* and *Less Established* for both locations.

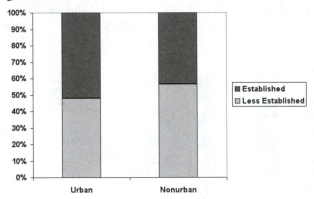

49. Worldwide toy sales.

   a) Row percentages (add up to 100%).

   b) No. We are given only the conditional distributions. We have no idea how much are sold in either Europe or America.

c)

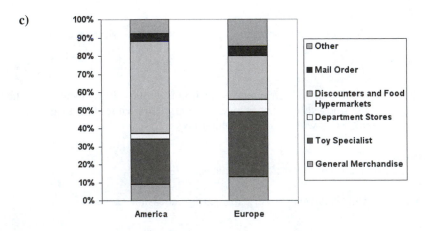

d)   In America, more than 50% of all toys are sold by large mass merchant discounters and food hypermarkets and only 25% are sold in toy specialty stores. In Europe, 36% of all toys are sold in toy specialty stores while a relatively small 24% are sold in the large discount and hypermarket chains.

51. Insurance company, part 2.

a)   The marginal totals were added. 160 of 1300 or 12.3% had a delayed discharge.

	Large Hospital	Small Hospital	Total
Major surgery	120 of 800	10 of 50	**130 of 850**
Minor surgery	10 of 200	20 of 250	**30 of 450**
Total	**130 of 1000**	**30 of 300**	**160 of 1300**

b)   Major surgery patients were delayed 15.3% of the time. Minor surgery patients were delayed 6.7% of the time.

c)   Large Hospital had a delay rate of 13%. Small Hospital had a delay rate of 10%. The small hospital has the lower overall rate of delayed discharge.

d)   Large Hospital: Major Surgery 15% and Minor Surgery 5%.
Small Hospital: Major Surgery 20% and Minor Surgery 8%.

e)   Yes, while the overall rate of delayed discharge is lower for the small hospital, the large hospital did better with *both* major and minor surgery.

f)   The small hospital performs a higher percentage of minor surgeries than major surgeries. 250 of 300 surgeries at the small hospital were minor (83%). Only 200 of the large hospital's 1000 surgeries were minor (20%). Minor surgery had a lower delay rate than major surgery (6.7% to 15.3%), so the small hospital's overall rate was artificially inflated. The larger hospital is the better hospital when comparing discharge delay rates.

53. Graduate admissions.

a)   1284 applicants were admitted out of a total of 3014 applicants. 1284/3014 = 42.6%

b)   1022 of 2165 (47.2%) of males were admitted. 262 of 849 (30.9%) of females were admitted.

c)   Since there are four comparisons to make, the table below organizes the percentages of males and females accepted in each program. Females are accepted at a higher rate in every program.

Program	Males Accepted (of applicants)	Females Accepted (of applicants)	Total
1	511 of 825	89 of 108	600 of 933
2	352 of 560	17 of 25	369 of 585
3	137 of 407	132 of 375	269 of 782
4	22 of 373	24 of 341	46 of 714
Total	**1022 of 2165**	**262 of 849**	**1284 of 3014**

Program	Males	Females
1	61.9%	82.4%
2	62.9%	68.0%
3	33.7%	35.2%
4	5.9%	7%

d)   The comparison of acceptance rate within each program is most valid. The overall percentage is an unfair average. It fails to take the different numbers of applicants and different acceptance rates of each program. Women tended to apply to the programs in which gaining acceptance was difficult for everyone. This is an example of Simpson's Paradox.

Chapter 5 – Displaying and Describing Quantitative Data

SECTION 5.1

   1.

       a)

       b)

       c)

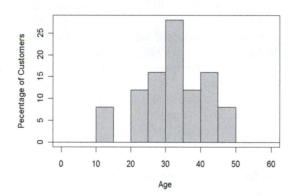

       d)   1 | 14
                2 | 0225699
                3 | 0002224558
                4 | 244488

SECTION 5.2
  3.

      a)   The distribution is unimodal.

      b)   The mode is around 35 years old.

      c)   The distribution is fairly symmetric.

      d)   No outliers are evident.

SECTION 5.3
  5.

      a)   The mean and median age should be about the same because the distribution is fairly symmetric.

      b)   31.84 years

      c)   32 years

SECTION 5.4
  7.

      a)   Q1= 26 ; Q3= 38 (answers may vary slightly)

      b)   Q1= 26; Q3= 38

      c)   IQR = 12 years

      d)   Standard Deviation = 9.84 years

SECTION 5.5
  9.

      a)   Shape – the distribution is clearly skewed to the right. Center – it is more difficult to determine visually the center of a skewed distribution. The center of a skewed distribution is best represented by the median, the exact middle data point when the data set is ordered either in ascending or descending order. There are 5000 data points representing the 5000 charge customers. The center of the data would be just to the right of the 2500th data point. It can be estimated which bin contains the median value by adding up the values in each bin. For example, the first bin (-$1000 to -$500) contains about 10 data points. The next bin ((-$500 to $0) contains slightly more, about 15 data points (the exact number is not important in this estimation of the median value) and the next bin ($0 to $5000) contains about 810 values. The next bin ($5000 to $10,000) contains about 720 values. The next bin ($1000 to $1500) contains about 830 values. The total so far is 10+15+810+720+830 = 2385 values. The 2500th data point has not yet been reached. The $1500 to $2000 bin contains a large number of values (about 750 values) which means the 2500th value is contained in that interval. Therefore, the center of the distribution is estimated to be between $1500 and $2000. Spread – the spread is determined from the range of data, low to high, or $5000-(-$1000) = approximately $6000. The exact range cannot be determined from the histogram because the intervals or bins do not represent the exact data points. There are no outliers. Unusual features – it can be pointed out that there are a few negative values that represent customers that received more credits than charges in the month; therefore the charge shows up as negative on the histogram.

      b)   The mean will be larger than the median because the distribution is right skewed. The median represents the exact middle number whereas the mean is an average of all data points, including the data points with higher values represented in the right tail. The mean will be pulled toward the tail with the higher values. The median is always the center value whether the distribution is symmetric or skewed.

      c)   The median is a more appropriate measure of the center of the distribution when it is skewed because it represents the middle or more typical value. The mean has been pulled toward the right or higher end because of the skew and therefore is not an accurate representation of the center.

SECTION 5.6
  11.

      a)

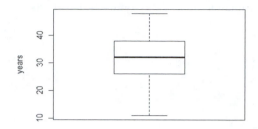

      b)   No outliers are nominated on the boxplot.

      c)   32+1.5*12 = 50 years old

  13.

      a)   The distribution can be described as skewed to the right. Symmetry can be determined by comparing the mean and the median. The mean is 46.50 and the median is 33.50. The mean is much larger than the median indicating a right skew (the higher values are pulling the mean value higher than the median). In addition, symmetry can be determined by comparing the difference between the first quartile and the median and the third quartile and the median. If the distribution is symmetric, these values should be fairly equal. In this summary, the median – Q1 = 33.50 – 18.50 = 15 compared to Q3 – median = 55 – 33.50 = 21.5. The right side of the distribution is wider than the left which indicates a right skew. Finally, the maximum value of 250 is very high compared to the median of 33.50 while the minimum value of 6 compared to the median of 33.50 is a much smaller number also indicating a skew to the right.

      b)   Yes, there is one high outlier at 250. This is clearly shown in the histogram from Exercise 10. In addition, its z-score of 4.26 indicates that 250 is more than 4 standard deviations above the mean.

      c)   The boxplot shows the outlier at 250 but without the data set, the length of the upper whisker going to the upper fence (limit for the outliers) cannot be determined.

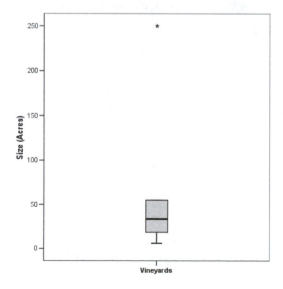

SECTION 5.7

15. The ages of the women are generally higher than men by about 10 years. More than ¾ of the women are older than all of the men. Both distributions are roughly symmetric, no outliers are evident.

17. In describing side-by-side boxplots, there are two important things to mention – the description of how each data set is distributed and a comparison of the distributions with each other. The distribution of weekly sales at Location #1 are roughly symmetric with some high outliers, one exceeding $320,000. The median sales value is approximately $240,000 and the minimum sales value is approximately $160,000. The distribution of weekly sales at Location #2 is also roughly symmetric with high outliers close to $150,000 to $180,000. The median sales value is approximately $110,000 and the minimum sales value is below $100,000.   Location #1 clearly has higher sales than Location #2 in every week except for the high outlier in Location #2.

SECTION 5.9

19.

a) Calculate the z-score for the minimum age of 11 as $z = \dfrac{11-31.84}{9.84} = -2.12$.

Calculate the z-score for the maximum age of 48 as $z = \dfrac{48-31.84}{9.84} = +1.64$.

Therefore, 11 has a z-score of -2.12 ; 48 has a z-score of +1.64.

b) The minimum age of 11 is more extreme because it is further away from the mean. It is more than 2 standard deviations below the mean while the maximum age is 1.64 standard deviations above the mean.

c) Solve for x in the following: $3 = \dfrac{x-31.84}{9.84}$

$x = 61.36$ years old.

SECTION 5.11

21.

The original values were: 16, 16, 36, 49, 49, 64, and 100.
The distribution of these values is skewed to the high end.

CHAPTER EXERCISES

23. **Statistics in business.** Answers will vary.

25. **Two-year college tuition.** Shape – the distribution is approximately symmetric with a single peak, making it unimodal. Center – the center of any distribution not perfectly symmetric is best represented by the median, the exact middle data point when the data set is ordered either in ascending or descending order. For a data set that is approximately symmetric with a clear peak, the center can be identified visually as being in the interval represented by the peak, in this case, the interval for $2-3000. The distribution is centered in the interval between 2000 and 3000 so it can be approximated at $2500. Spread – the spread is determined from the range of data, low to high, or $6000-$0 = approximately $6000. The exact range cannot be determined from the histogram because the intervals or bins do not represent the exact data points. There are no outliers or other unusual features in this distribution. It can be pointed out that most of the tuitions lie between $1000 and $4000, which includes 45 out of 50 states.

27. **Mutual funds.** Shape – the distribution is approximately symmetric with a single peak, making it unimodal. Center – the center of the distribution is close to the single peak but to be sure, the exact number of values and the median should be determined. The bin values can be added up as follows: $1+3+0+7+18+14+8+4+3+2+2+1+1 = 64$. The median occurs to the right of the 32nd data point. Twenty nine data points are contained in the first five bins. The 32nd data point is therefore contained in the 6th bin representing approximately 10%. Spread – the spread is determined from the range of data, high minus low, or 90%-(-15%) = approximately 105%. The two highest bins at 80% and 85% contain outliers.

29. **Mutual funds, part 2.**

    a)   Five-Number Summary   (Quartile calculations may differ slightly using different software)

Min	1st Qtr	Median	3rd Qtr	Max
−10.820	6.965	11.275	17.440	94.940

    b)   Center can be represented by the median which is 11.275%. The spread can be represented by the maximum – minimum values which is 94.940 – (−10.820) = 105.76%. The IQR (3rd quartile – 1st quartile values) summarizes the spread by focusing on the middle 50% of the data. For this data set, the IQR = 17.440 – 6.965 = 10.475%. Another measure of spread is the standard deviation but this measure is reserved for symmetric distributions. The shape of the distribution can be determined from the histogram in Problem 7. The main distribution is fairly symmetric but there are high outliers, meaning that it is not appropriate to use standard deviation as a measure of spread. In addition, due to the outliers, it is also not accurate to represent the data by a simple calculation of maximum – minimum values. For this distribution, it is better to represent the spread using the IQR value.

    c)   Boxplot

    d)   The histogram clearly shows the outlier values as well as the skewness. Depending on the software package used to create a boxplot, the outliers may be identified using special symbols as shown above.

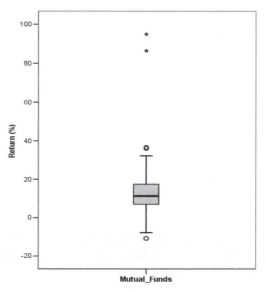

31. Vineyards. The stem and leaf plot has intervals of 20 acres (the 0 interval includes single digits and teens up to but not including 20). The distribution is shown to be clearly right skewed as seen before in the boxplot and histogram. Most of the data points end in either 0 or 5 indicating that the measurements may have been rounded.

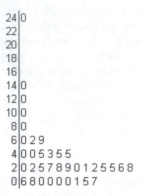

```
24|0
22|
20|
18|
16|
14|0
12|0
10|0
 8|0
 6|0 2 9
 4|0 0 5 3 5 5
 2|0 2 5 7 8 9 0 1 2 5 5 6 8
 0|6 8 0 0 0 0 1 5 7
```

33. Gretzky.

    a) Stemplot - The stem and leaf plot has split stems representing 0-4 and 5-9.

```
8|
8|0 0 0 0 0 0 1 2 2
7|8 8 9 9
7|0 3 4 4
6|
6|4
5|
5|
4|5 8
4|
```

    b) Boxplot

    c) The distribution of the number of games played per season by Wayne Gretzky is skewed to the left toward lower values and has low outliers. The median value is 79 games and the range is 37 games.

    d) The outliers identified at 45 and 48 games could have been caused by injuries. The season with 64 games has a gap to both lower and higher values. Most of his seasons were played with games totaling in the 70s and 80s.

35. Gretzky returns.

  a) The distribution is skewed therefore the median is used to describe the center.

  b) The mean should be pulled toward the tail of the distribution, in this case, toward the lower values.

  c) The chart displayed is not a histogram. It is a time series plot using bars to represent each data point. A histogram would arrange the data into bin intervals rather than displaying the number of games over time.

37. Pizza prices.

  a) Dallas Five-Number Summary (Quartile calculations may differ slightly using different software)

Min	1st Qtr	Median	3rd Qtr	Max
$2.21	$2.51	$2.61	$2.72	$3.05

  b) Range = max – min = $3.05 - $2.21 = $0.84; IQR = Q3 – Q1 = $2.72 - $2.51 = $0.21

  c) Boxplot

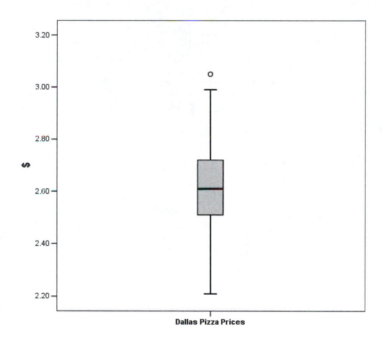

  d) This distribution is fairly symmetric with a high outlier at $3.05. The mean is $2.62 and the standard deviation is $0.156.

  e) There is one observation classified as an outlier at $3.05 per frozen pizza.

39. Gasoline usage. A report for a given data set should include summaries and graphs with analysis. Five-Number Summary (Quartile calculations may differ slightly using different software)

Min	1st Qtr	Median	3rd Qtr	Max
209.47	429.168	485.73	510.918	589.18

Range = max – min = $589.18 - $209.47 = $379.71; IQR = Q3 – Q1 = $510.918 – $429.168 = $81.75

The distribution of gas usage is unimodal (single peak) and skewed to the left. Two outliers are identified that represent the District of Columbia and New York state. D.C. is a city and New York has a large population in New York City. Public transportation in those cities could affect the gas usage. The median usage is 485.73 gallons per year per capita. The IQR is 81.75 gallons/yr. Minimum usage is in D.C. with 209.47 gallons and the maximum is Wyoming with 589.18 gallons.

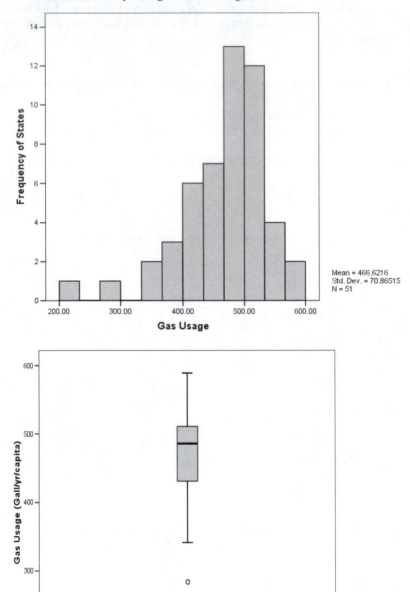

41. Golf courses.

    a)   Range: max – min = 6796 – 5185 = 1611 yards

    b)   The middle 50% of the distribution lies between Quartile 1 (5585.75 yards) and Quartile 3 (6131 yards) (Quartile calculations may differ slightly using different software).

c) Summary statistics should include the representation of the center, in this case, the mean (5893 yards) because the distribution is approximately symmetric. In addition, a measure of the spread for this data set is the standard deviation (386.6 yards), chosen also because the distribution is approximately symmetric.

d) Shape – the distribution of the lengths of all of the Vermont golf courses is roughly symmetric and unimodal. Center – the mean is 5893 yards, approximately 5900 yards. Spread – represented by the standard deviation of 386.6 yards.

43. Food sales.

a) Suitable displays of a single quantitative variable are either a histogram or a boxplot. Both are shown here:

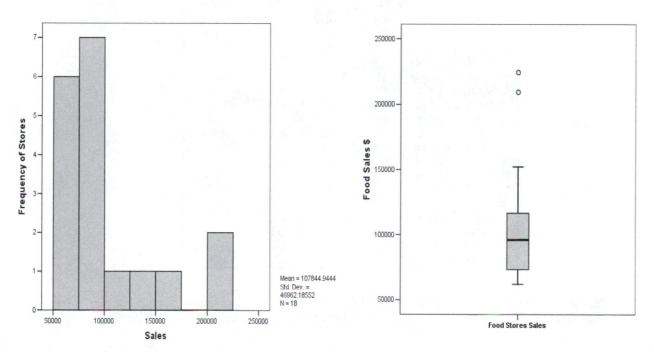

Mean = 107844.9444
Std. Dev. = 46962.18552
N = 18

b) The median for the data set is $95,974.5 and the mean is $107,844.94. The mean is pulled toward the higher values because the distribution is right skewed and the higher values have the effect of increasing the mean value.

Summary of Sales	
Count	18
Mean	107844.944
Median	95974.5
Std Dev	46962.186
Variance	2205446869.114
Range	162498
Min	62006
Max	224504
IQR	43300
25th%	73320
75th%	116620

c) The median does a better job of depicting typical store sales because the distribution is skewed with high outliers.

d) Standard Deviation = $46,962.19 and the IQR = $43,300 (Quartile calculations may differ slightly using different software)

e) The IQR is a better measure of the spread for a skewed distribution with outliers. The standard deviation is affected by outliers.

f) If the outliers were removed, the mean would decrease in value because the higher numbers would not be included in the calculation. The standard deviation would also decrease, indicating a smaller spread when the high values are excluded. The median and IQR would remain relatively unaffected because the calculations are not affected by outliers unless there are a large number of them.

45. IPod failures. The failure rate is calculated by dividing the number failed by the total number of iPods (both failed and OK) for each model. The result is then multiplied by 100% to get a rate.

	A	B	C	D	E
1	Product	OK	Failed	Total	Failure Rate
2	5GB Scroll Whl	784	234	1018	23.0%
3	10GB Scroll Whl	212	58	270	21.5%
4	10GB Touch Whl	270	54	324	16.7%
5	20GB Touch Whl	339	58	397	14.6%
6	10GB Dock Cntr	160	25	185	13.5%
7	15GB Dock Cntr	342	92	434	21.2%
8	30GB Dock Cntr	244	77	321	24.0%
9	20GB Dock Cntr	397	68	465	14.6%
10	40GB Dock Cntr	338	84	422	19.9%
11	20GB Click Whl	512	129	641	20.1%
12	40GB Click Whl	289	123	412	29.9%
13	40GB Photo	181	35	216	16.2%
14	60GB Photo	272	29	301	9.6%
15	with Color 20GB	142	10	152	6.6%
16	with Color 60GB	82	10	92	10.9%
17	30GB Video	135	6	141	4.3%
18	60GB Video	183	6	189	3.2%

The median value for the failure rate for all 17 models is 16.2%. An appropriate graphical display of the distribution of a single quantitative variable is either a histogram or a boxplot.

▷ Summary of Failure Ra…	
Count	17
Mean	0.159
Median	0.162
Std Dev	0.0734
Variance	0.00539
Range	0.267
Min	0.0317
Max	0.299
IQR	0.107
25th%	0.106
75th%	0.213

The distribution is left skewed. The center is best represented by the median at 16.2% and the spread is best represented by the IQR which is 10.7% (Quartile calculations may differ slightly using different software). The middle 50% of failure rates are between 10.6% and 21.3%. The lowest value or best failure rate is 3.2% for the 60GB Video model, and the worst failure rate is 29.9% for the 40GB Click Wheel.

47. Gas prices, part 2.

    a)   It is obvious that gas prices have steadily increased over the three-year period from 2002 – 2004. The data spread has also increased over these three years. The 2002 distribution of prices is skewed to the left with several low outliers. Starting in 2003, the distributions have become increasingly skewed to the right. There is a high outlier in 2004 which is close to the upper fence.

    b)   The evidence for prices showing more volatility would be prices showing a greater spread and a larger IQR. The year with the greatest spread and IQR is 2004.

49. Wine prices.

    a)   Identify the highest value for the three regions – occurs in Seneca Lake.

    b)   Identify the lowest value for the three regions – occurs in Seneca Lake.

    c)   To answer the question about which wines are generally more expensive, look at the IQR box and determine which region has the middle 50% of its prices higher than the others. That region is clearly Keuka Lake.

    d)   Cayuga Lake and Seneca Lake vineyards have approximately the same price at about $200 per case. The middle 50% of prices for these two regions are also similar, from approximately $150 to $220 per case. A typical Keuka Lake vineyard case has a much higher price of about $260 and the middle 50% values are between $240 and $280 per case with one outlier at approximately $170 per case. Seneca Lake vineyard prices are the most varied and include both the lowest and the highest prices for all three regions (from $100 to $300).

51. Derby speeds.

   a)   The median speed is the speed at the middle of the data values where 50% run slower and 50% run faster. The values represent the percentage of Kentucky Derby winners that have run slower than the cutoff of 30 mph. The data set already represents a percentage of the total distribution so the median and quartiles can be determined by looking up the percentage values on the y-axis. The 50% value can be determined at the 50% mark on the y-axis and moving over to the plotted points to approximately 36 mph (find 50% on the y-axis, move straight over to the right and identify the value on the plot on the x-axis).

   b)   The quartiles are found in a similar way. Find the winning speed representing 25% for the first quartile: Q1 = approximately 34.5 mph. Find the winning speed representing 75% for the third quartile: Q3 = approximately 36.5 mph.

   c)   Range = values that represent 0% to 100% = 31 to 38 mph = 7 mph. The IQR is 75% value – 25% value = 36.5 mph – 34.5 mph = 2 mph.

   d)   Boxplot.

   e)   The distribution of speeds is skewed to the left. The lowest speed is close to 31 mph and the fastest speed is 37.5 mph. The median speed is approximately 36 mph. Twenty five percent of the speeds are above 36.5 mph and 75% of winning speeds are above 34.5 mph. Only a small percent of winners had speeds below 33 mph. Without the actual data set, the boxplot is constructed from the five # summary and therefore fences and outliers other than the maximum and minimum cannot be determined.

53. Test scores.

   a)   The highest mean score can be determined from looking at the shape of the test histograms. Class 1 is symmetric with a mean in the 60 interval. Class 2 is somewhat symmetric with the center also being in the 60 interval. Class 3 is left skewed with most of the data points to the right of a score of 60 so the mean for Class 3 would be the highest.

   b)   The same logic can be applied to find the median score. For symmetric distributions, the mean is the same as the median and close to the median for approximately symmetric distributions. Most of the data points for Class 3 are to the right of a score of 60 therefore the median for Class 3 would be the highest.

   c)   For symmetric and approximately symmetric distributions, the mean and the median are almost the same. That applies to both Class 1 and Class 2. Class 3 will have a mean pulled toward its distribution tail which is to the left for a left skewed distribution. Therefore, the difference between the mean and the median will be greatest for Class 3.

d)   The smallest standard deviation represents the smallest data spread from the mean of the distribution. Because Class 1 is symmetric and clustered more tightly around the center (about 60), it will have the smallest standard deviation.

e)   The smallest IQR represents the most condensed data in the middle 50% region. Class 1 probably has the smallest IQR because it is the most symmetric, however, without the actual scores, it is impossible to calculate the exact IQRs.

55. Quality control holes. In order to analyze the data, it is appropriate to create summaries separately for Fast and Slow data sets. In addition, create side-by-side boxplots for comparison of distributions.

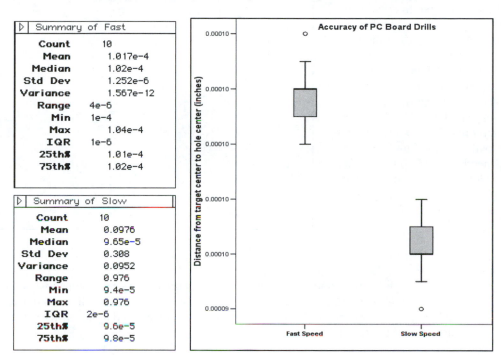

The slow drilling data contains an extremely high outlier indicating that one hole was drilled almost an inch away from the center of the target. If this data point is correct, the engineers should investigate the slow speed drilling process closely for any extreme, intermittent inaccuracy. The outlier is so extreme that no graphical display can show the distributions in a meaningful way if that data point is included. The outlier can be removed in order to look at the remaining data points. However, it should be noted that it is never good scientific practice to eliminate a data point because it doesn't fit the distribution. All outliers should be investigated to determine if they are a result of human error, input error, or some problems with the measurements that need to be addressed. It seems apparent that the entry of .9756000 was in error and was meant to be .000009576. This should be investigated and if found to be true, the summary statistics should be recalculated.

But with the outlier removed, the slow drilling process is shown to be more accurate. The entire distribution with the exception of the extreme outlier not pictured lies well below the fast distribution. The greatest distance from the target for the slow drilling process is 0.000098 inches which is more accurate than the smallest distance for the fast drilling process, 0.000100 inches.

57. Customer database.

   a)   The mean of 54.41 is meaningless. The data set is categorical, not quantitative.

   b)   Typically, the mean and standard deviation are influenced by outliers and skewness, however, once again, these results are meaningless in this instance because the values are categorical.

   c)   No. Summary statistics are only appropriate for quantitative data.

59. Mutual funds types. Over the 3-month period, International Funds generally outperformed the other two types of mutual funds. Almost half of the International Funds outperformed all the funds in the other two categories. U.S. Domestic Large Cap Funds did better than U.S. Domestic Small/Mid Cap Funds in general. Large Cap funds had the least variation of the three types.

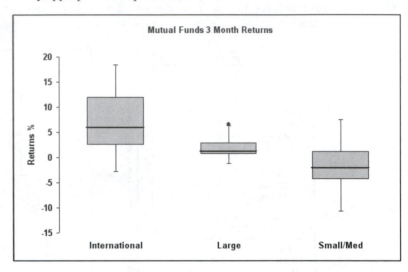

61. Houses for sale.

   a)   Even though MLS ID numbers are categorical identifiers, they are assigned sequentially, so this graph can be analyzed as shown. Most if not almost all of the houses listed a long time ago have sold and are no longer listed. The older numbers are represented by the lower values giving the distribution the left skewed shape.

   b)   Although some information could be gathered from this graph, a histogram is generally not an appropriate display for categorical data. The MLS ID numbers are categorical identifiers.

63. Hurricanes.

   a)   Histogram

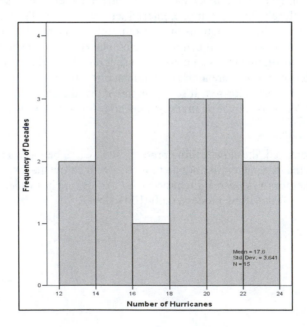

b) The distribution is fairly uniform and appears somewhat right skewed. There do not appear to be any outliers.

c)

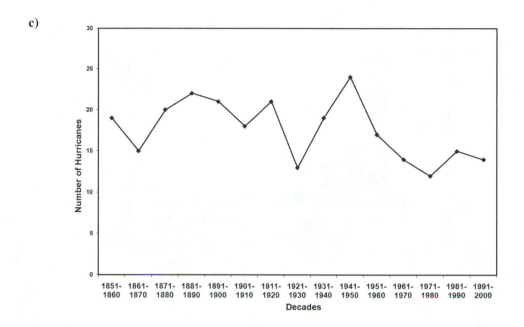

d) The timeplot does not support the claim that the number of hurricanes has increased in recent decades. There was a peak in the 1940's but numbers have decreased since that time.

65. Productivity study. Questions include: What is plotted on the x-axis? If it is time, what are the units? Months? Years? Decades? How is productivity measured?

67. Real estate, part 2. To answer this question, the values have to be standardized and the z values compared according to the formula: $z = \dfrac{(y - \overline{y})}{s}$ where y is the value being compared to $\overline{y}$ which is the mean value and s is the standard deviation. The house that sells for $400,000 has a z-score of $(400000 - 167900)/77158 = 3.01$. The house with 4000 sq. ft. of living space has a z-score of $(4000 - 1819)/663 = 3.29$. The value of 3.29 is further away from the center of a normal distribution than 3.01 and, therefore, is the more unusual value.

69. Food consumption.
To answer this question, the values have to be standardized and the z values compared according the formula: $z = \dfrac{(y - \overline{y})}{s}$ where y is the value being compared to $\overline{y}$ which is the mean value and s is the standard deviation. The mean and standard deviation need to be calculated from the given data set. For meat consumption, the U.S. z-score = $(267.30 - 181.031)/53.077 = 1.625$ and the Ireland z-score = $(194.26 - 181.031)/53.077 = 0.25$. Therefore, the U.S. has a more remarkable meat consumption than Ireland because the z-score is higher, indicating further from the distribution mean.

To answer this question, find the z-scores separately for meat and alcohol for each country. For meat consumption, the U.S. z-score = 1.625 and the Ireland z-score = 0.25. For alcohol consumption, the U.S. z-score = $(26.36 - 26.778)/10.469 = -0.04$ and the Ireland z-score = $(55.80 - 26.778)/10.469 = 2.77$. The total for the U.S. is $1.625 + (-0.04) = 1.585$. The total for Ireland is $0.25 + 2.77 = 3.02$. Ireland has higher overall consumption for meat and alcohol combined.

71. Regular gas.

    a)   The histogram of gas prices is strongly skewed to the right. Prices range from slightly less than $1 to slightly over $3/gallon for the time period measured.

    b)   The time series plot shows that prices remained relatively stable during the early and mid 1990's and then began increasing significantly around 1999 until 2005.

    c)   The time series plot is more appropriate to show the how gas prices have changes over the past several years.

73. Unemployment rate, 2010.

    a)   The histogram shows that the distribution is possibly skewed to the left and bimodal.

    b)   The trend over time.

    c)   The time series plot because it demonstrates more of the structure of the data, how the data change over time.

    d)   Unemployment decreased steadily from approximately 5.6% in 1995 to below 4.0% by the year 2000. Then the rate increased sharply over the next year and finally rose to between 5.5 and 6.4% in the 2002 to 2004 timeframe.

75. Assets.

    a)   The distribution is right skewed. That makes it difficult to estimate the mean which is pulled toward the right. The median can be determined from previous methods: there are 79 total data points and the median is located at the 40th data point. The first interval (0–$4000) contains over 40 data points and therefore contains the median value. The spread for a skewed distribution cannot be easily determined.

    b)   It might be possible to transform the data using logs or square roots which can make the data more symmetric.

Chapter 6 – Correlation and Linear Regression

SECTION EXERCISES

SECTION 6.1
1.

    a)   The scatterplot of *Sales* against *Number of sales people working* is shown below.

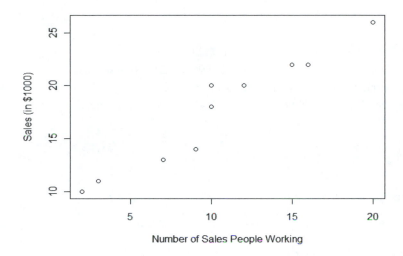

    b)   The direction of the association is positive. *Sales* increases as the *Number of sales people working* increases.

    c)   The form of the relationship is linear.

    d)   Since the points appear tightly clustered along the line, the strength of the relationship is strong.

    e)   There is no outlier evident in the scatterplot.

SECTION 6.2
3.

    a)   The explanatory (predictor) variable is *Years of experience*.

    b)   The response variable is *Salary*.

    c)   *Salary* would be plotted on the *y* axis.

SECTION 6.3
5.

    a)   True

    b)   False. Correlation is not affected by changes in the center or scale of either variable; consequently multiplying every value of *x* will not change the correlation.

    c)   False. As a standardized measure, the correlation has no units.

SECTION 6.4
7.   Correlation does not necessarily imply causation. The analyst's argument is that the *Number of sales people working* causes *Sales* to increase. There may very well be a lurking variable. For example, the store may hire more sales staff when they anticipate an increase in sales, for example when there is a book signing event or after a promotional campaign.

SECTION 6.5

9.

    a) False. We choose the best fitting line through the data points on a scatterplot. This is done by minimizing the sum of the squared errors. The predicted values (not the data values) fall on the line.

    b) True

    c) False. Least squares means that the sum of all squared residuals (difference between the observed y values and the y values predicted by the line) is minimized.

SECTION 6.6

11.

    a) We would expect sales to be 2 x 0.965 = 1.93 SDs above the mean.

    b) This corresponds to 17.6 + 1.93 x 5.34 = 27.906 or $27,906 in sales.

    c) We would expect sales to be 0.965 SDs below the mean.

    d) This corresponds to 17.6 - 0.965 x 5.34 = 12.447 or $12,447 in sales.

13.

    a) $b_1 = 0.914$

        To find the slope use $b_1 = r\dfrac{s_y}{s_x}$

$$b_1 = 0.965 \times \frac{5.34}{5.64} = 0.914$$

    b) The slope of 0.914 indicates that an additional 0.914 ($1000) or $914 of sales is associated with each additional sales person working.

    c) $b_0 = 8.09$

        To find the intercept use $b_0 = \bar{y} - b_1\bar{x}$

$$b_0 = 17.6 - (0.914)(10.4) = 8.09$$

    d) The intercept of 8.09 implies that, on average, we expect sales of 8.09 ($1000) or $8090 with 0 sales people working. This value is not meaningful in this context.

    e) The equation that predicts *Sales* from *Number of Sales People Working is*
$\hat{y} = 8.09 + 0.914x.$

    f) Using the equation to predict sales if 18 people are working, we have
$\hat{y} = 8.09 + 0.914(18) = 24.542$ or 24.542($1000) = $24,542

    g) The value of the residual is 25-24.542 = 0.458 ($1000) or $458.

    h) Because the actual sales value is more than the predicted value, the estimated regression equation has underestimated sales when $x = 18$.

SECTION 6.7

15. The winners may be suffering from regression to the mean. Regression toward the mean implies that a variable which yields an extreme value on the first measurement will tend to be closer to the mean on a later measurement. So the winners of the rookie junior executive of the year awards had an extreme measurement their first year. This extremely good performance may have been due, in part, to just an unusually lucky year. Their next year's performance moved closer to the mean.

SECTION 6.8

17.

    a) The units of the residuals are the same as the units of the response variable, *Sales.* Therefore the units are thousands of dollars.

    b) The residual with the largest absolute value contributes the most to the sum being minimized by the Least Squares Criterion. In this case it is 2.77.

    c) The residual with the smallest absolute value contributes the least to the sum being minimized by the Least Squares Criterion. In this case it is 0.07.

SECTION 6.9

19. $R^2 = 93.12\%$. R^2 is the squared correlation, or $.965^2$. It means that about 93% of the variance in *Sales* can be accounted for by the regression model of *Sales* on *Number of Sales Workers.*

SECTION 6.11

21.

    a) The original values are the squared resulting values: 16, 16, 36, 49, 49, 64, and 100.

    b) The values are not symmetric but instead skewed to the right (high end).

CHAPTER EXERCISES

23. Association.

    a) The number of text messages sent is the explanatory variable and the cost is the response variable. Cost is predicted from the number of text messages sent. The association is positive, linear, and moderately strong. The effect of increased text messages will be increased costs. A possible outlier or data that might not fit the trend could be the result of cell phone contracts with fixed costs for texting.

    b) Fuel efficiency is the explanatory variable and the sales volume is the response variable. In a time of high fuel costs, cars with better fuel efficiency should sell more (a positive association) but due to the higher costs of some of these vehicles, they may not sell during hard economic times. Therefore, we cannot say with certainty what the association is and we have no information about the shape or strength of the relationship.

    c) At first, it appears that there should be no association between ice cream sales and air conditioner sales. When the lurking variable of temperature is considered, the association becomes more apparent. When the temperature is high, ice cream sales tend to increase. Also, when the temperature is high, air conditioner sales tend to increase. Therefore, there is likely to be an increase in the sales of air conditioners whenever there is an increase in the sales of ice cream. The association would be positive, possibly linear, with moderate strength. Either one of the variables could be used as the explanatory variable.

    d) Price is the explanatory variable and demand (number sold per day) is the response variable. Demand is predicted from price. The association would have a negative direction. It would be a linear shape in a narrow range but would be curved over a larger range of prices.

25. Scatterplots.

    a) None

    b) #3 and #4 show a negative association. Increases in one variable are generally related to decreases in the other variable.

    c) #2, #3, and #4 show at least some straight line association.

    d) #2 shows a moderately strong association.

    e) #3 and #1 show a strong association. #3 is linear and #1 is non-linear.

27. Manufacturing.

    a) Histogram:

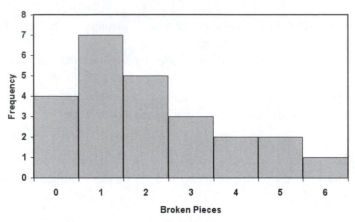

    b) The distribution is unimodal and skewed to the right. The skewness is more apparent in the histogram.

    c) The positive, somewhat linear relationship between batch number and broken pieces is more apparent in the scatterplot.

29. Matching.

    a) 0.006 (relationship is curved so linear correlation will be almost zero)

    b) 0.777 (positive, moderately strong)

    c) -0.923 (negative, very strong)

    d) -0.487 (negative, moderately weak)

31. Pizza sales and price.

    a) The problem states that the weekly *Sales* (in pounds) of frozen pizza are being predicted from the average *Price*/unit ($). The variable being predicted is the response variable and the variable doing the predicting or explaining is the explanatory variable. In this case, the explanatory variable is the unit *Price* (pizza *Sales* predicted from unit *Price*). In addition, the prediction equation is stated as: $\widehat{Sales} = 141{,}865.53 - 24{,}369.49 * Price$ which is of the form: $\hat{y} = b_0 + b_1 x$. *Price* is x, or the explanatory variable.

    b) The problem states that the weekly *Sales* (in pounds) of frozen pizza are being predicted from the average *Price*/unit ($). Whatever is being predicted is the response variable. *Sales* of frozen pizza is being predicted and, therefore, *Sales* is y or the response variable. In addition, the prediction equation is stated as: $\widehat{Sales} = 141{,}865.53 - 24{,}369.49 * Price$ which is of the form: $\hat{y} = b_0 + b_1 x$. *Sales* is y, or the response variable.

    c) The prediction equation is stated as: $\widehat{Sales} = 141{,}865.53 - 24{,}369.49 * Price$ which is of the form: $\hat{y} = b_0 + b_1 x$ where b_1 is the slope. The slope for this equation is –$24,369.49, which means that for every extra dollar increase in price, weekly sales of frozen pizzas is predicted to decrease by 24,369.49 pounds.

d) The y intercept is the value of the line when the x variable is zero. The intercept can be used as a starting point for predictions but it is not meaningful in all circumstances. In this equation, $\widehat{Sales} = 141{,}865.53 - 24{,}369.49 * Price$ is of the form: $\hat{y} = b_0 + b_1 x$ where b_0 is the y intercept. The y intercept for this equation is \$141,865.53. This number is not meaningful except as a base or starting value for the line because it is obviously not realistic to set the *Price* at zero dollars.

e) A prediction can be made by substituting the given value for the *Price* of pizza into the equation and solving for weekly *Sales*. If the *Price*/unit of the pizza is \$3.50, the equation is:

$\widehat{Sales} = 141{,}865.53 - 24{,}369.49 * (3.50) = 141{,}865.53 - 85{,}293.22 = 56{,}572.32$ pounds.

f) If a sales *Price* of \$3.50 yields 60,000 pounds, the residual is calculated by subtracting the predicted value (calculated from the linear model equation) from the observed or measured value (*Residual = Data – Predicted* or $e = y - \hat{y}$). The predicted value of y at an x value of \$3.50 was calculated in part e. to be 56,572.32 pounds. The *Residual* = 60,000 (*Data*) –56,572.32 (*Predicted*) = 3,427.68 pounds.

33. Football salaries.

a) The problem states that NFL *Wins* (out of 16 regular season games) are being predicted from the total team *Salary* (in \$M). The variable being predicted is the response variable and the variable doing the predicting or explaining is the explanatory variable. In this case, the explanatory variable is the team *Salary* (*Wins* predicted from *Salary*). In addition, the prediction equation is stated as:

$\widehat{Wins} = 1.783 + 0.062\,Salary$ which is of the form: $\hat{y} = b_0 + b_1 x$. *Salary* is x, or the explanatory variable.

b) The problem states that *Wins* (out of 16 regular season games) are being predicted from the total team *Salary* (in \$M). The variable being predicted is the response variable and the variable doing the predicting or explaining is the explanatory variable. *Wins* are being predicted and, therefore, *Wins* is y or the response variable. In addition, the prediction equation is stated as:

$\widehat{Wins} = 1.783 + 0.062\,Salary$ which is of the form: $\hat{y} = b_0 + b_1 x$. *Wins* is y, or the response variable.

c) The slope represents the change in y or the response variable for every x unit or one unit step in the predictor variable. The prediction equation is stated as: $\widehat{Wins} = 1.783 + 0.062\,Salary$ which is of the form: $\hat{y} = b_0 + b_1 x$ where b_1 is the slope. The slope for this equation is 0.062, which means that for every one million increase in *Salary*, the number of *Wins* increases by 0.062 (for a \$10M increase in total *Salary*, number of *Wins* increases by 0.62 – not even a full game).

d) The y intercept is the value of the line when the x variable is zero. The intercept can be used as a starting point for predictions but it is not meaningful in all circumstances. In this equation, $\widehat{Wins} = 1.783 + 0.062\,Salary$ is of the form: $\hat{y} = b_0 + b_1 x$ where b_0 is the y intercept. The y intercept for this equation is 1.783. This number represents the base or starting value for the line which means the number of season *Wins* for a team salary of \$0 is 1.783 games. This is not realistic.

e) If a team spends \$10M extra on their total salary, their Wins increases by 0.62 games (this calculation can be done by looking at the slope from part c.).

f) A team spends \$50M on salary and won 8 games. The predicted value of y at an x value of \$50M is: $\widehat{Wins} = 1.783 + 0.062\,Salary = 1.783 + 0.062 * 50 = 4.883$ The predicted value using the linear model is lower than the actual number of wins so they performed better than predicted.

g) If a team spends $50M on salary and won 8 games, the residual is calculated by subtracting the predicted value (calculated from the linear model equation) from the observed or measured value ($Residual = Data - Predicted$ or $e = y - \hat{y}$). The predicted value of y at an x value of $50M is

$$\widehat{Wins} = 1.783 + 0.062\, Salary = 1.783 + 0.062 * 50 = 4.883 \quad \text{The } Residual = 8\ (Data) - 4.883$$

($Predicted$) = 3.117 games.

35. Pizza sales and price, part 2.

$Average\ Sales = 141{,}865.53 - 24{,}369.49*(Average\ Price)$
$\$52{,}697 = \$141{,}865.53 - 24{,}369.49*(Average\ Price)$
Thus, $Average\ Price = (52{,}697 - 141{,}865.53)/(-24{,}369.49) = \$3.659$

Now, $b_1 = r * \left(\dfrac{s_y}{s_x}\right)$ or $-24{,}369.49 = (-0.547) * \left(\dfrac{10{,}261}{s_x}\right)$ or

$s_x = \left(\dfrac{(-0.547)*(10{,}261)}{-24{,}369.49}\right) = \$0.23031943$

So price one SD above the average = $\$3.659 + \$0.23031943 = \$3.88932$.
At that price, the Predicted Sales = $141{,}865.53 - 24{,}369.49*(3.88932) = 47{,}084.80$ pounds

37. Packaging. "Packaging" isn't a variable. It is more like a category. There is no basis for computing a correlation.

39. Sales by region. The model is meaningless because the variable Region is categorical, not quantitative. Although each region is denoted by a number, the variable is still categorical. The slope makes no sense because Region has no units. The boxplot comparisons are informative, but the regression is meaningless.

41. Carbon footprint.
    a) There is a strong negative linear association between *Carbon Footprint* and *Highway* mpg.

    b) The variables are quantitative and the relationship is straight (linear) enough. The Prius is far from the rest of the data. It is in line with the linear pattern but, due to being so removed, could be considered an outlier.

    c) Removing the Prius reduces the correlation slightly from -0.947 to -0.940. Data values far from the main body of the data and in line with the linear trend tend to increase correlation and could make the strength of the linear association misleading.

43. Real estate.
    a) There does appear to be a positive association between *Price* and the number of *Rooms*.

    b) The plot is not linear and violates the linearity condition. In addition, there may be an outlier at 17 rooms.

45. GDP growth.
    a) The variables are both quantitative, the trend is positive and somewhat straight, there are no outliers, and the spread is roughly consistent although the spread is large.

    b) About 21% of the variation in the growth rates of developing countries is accounted for by the growth rates of developed countries.

    c) Each point represents one of the years 1970–2007, which are the cases in the model.

47. GDP growth, part 2.

    a)   The output of regression analysis gives the coefficients of the linear model equation. The y intercept is given as 3.46. The slope, which is given as the coefficient of the explanatory variable (Annual GDP Growth of Developing Countries), is 0.433. Therefore, the linear model is:

$$\overline{Growth\,(Developing\,Countries)} = 3.46 + 0.433\,Growth\,(Developed\,Countries)$$

    b)   The y intercept is the value of the line when the x variable is zero. The intercept can be used as a starting point for predictions but it is not meaningful in all circumstances. In this equation,

$$\overline{Growth\,(Developing\,Countries)} = 3.46 + 0.433\,Growth\,(Developed\,Countries)\ \text{is of the form:}$$

       $\hat{y} = b_0 + b_1 x$ where b_0 is the y intercept. The y intercept for this equation is 3.46. This number represents the base or starting value for the line which means the Annual GDP Growth Rate of Developing Countries when the Annual GDP Growth Rate of Developed Countries is zero percent. This value is 3.46% and the concept makes sense.

    c)   The slope represents the change in y or the response variable for every x unit or one unit step in the predictor variable. The slope for this equation is 0.433, which means that for every 1% increase in *Annual GDP Growth of Developed Countries*, the *Annual GDP Growth of Developing Countries* increases by 0.4327%.

    d)   A prediction can be made by substituting the given value of 4% for *Annual GDP Growth of Developing Countries* into the equation and solving for *Annual GDP Growth of Developing Countries*. The equation is: $\overline{Growth\,(Developing\,Countries)} = 3.46 + 0.433 * 4$
$= 3.46 + 1.732 = 5.192\%$.

    e)   If developed countries experience a 2.65% growth while developing countries grew at a rate of 6.09%, the predicted value of y at an x value of 2.65% is:

       $\overline{Growth\,(Developing\,Countries)} = 3.46 + 0.433 * 2.65 = 4.60745\%$. The predicted value using the linear model is lower than the actual percentage. The actual value performed better than expected.

    f)   The residual is calculated by subtracting the predicted value (calculated from the linear model equation) from the observed or measured value (*Residual = Data – Predicted* or $e = y - \hat{y}$). The predicted value of y at an x value of 2.65% was calculated in e) as 4.60745%. The *Residual =* 6.09% (*Data*) – 4.60745%. (*Predicted*) = 1.48255% or close to 1.48%.

49. Attendance 2006.

    a)   The two variables are quantitative. The relationship between them appears to be straight with no outliers. Conditions for linearity are met.

    b)   The association between attendance and runs scored is positive, straight, and moderate in strength. Teams that score more runs generally have higher attendance.

    c)   There is evidence of an association between attendance and runs scored, but a cause-and-effect relationship between the two is not implied. There may be lurking variables that can account for the increases in each. For example, perhaps winning teams score more runs and also have higher attendance. We don't have any basis to make a claim of causation.

51. Tuition 2008.

   a)   The scatterplot shows a positive relationship that is moderately strong. There seems to be a flattening out or a slight downward bend on the upper tuition end. The data are somewhat spread out.

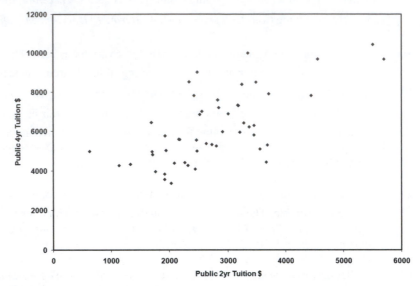

   b)   This relationship is expected. State funding affects both 4 year and 2 year colleges fairly equally. If the state decides that college education should cost more, they are more likely to increase tuition in both types of colleges.

   c)   The regression equation has to be calculated using a technology package and its output. Some packages print out the coefficients as shown in Exercises 9 and 10. Others place the regression line on the scatterplot and state the equation as shown below. The regression equation is of the form:

$\hat{y} = b_0 + b_1 x$. In this case, $\overline{Public\,4\,yr\,Tuition} = 2826 + 1.2157\,Public\,2\,yr\,Tuition$

   d)   A linear model does seem appropriate for this relationship. There are no major deviations.

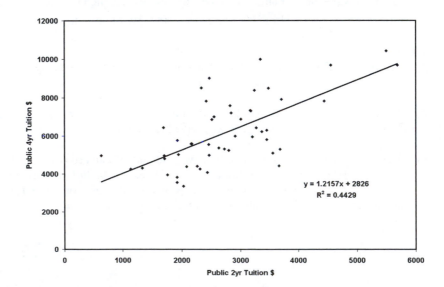

e)   This question can be answered by looking at the regression equation. *Public 4yr Tuition* will be $2826 plus 1.2157 (rounded to 1.216) * *Public 2Yr Tuition*. The y intercept gives the starting point for the y values (*Public 4yr Tuition*). In addition, the slope tells us that for every $1 increase in *Public 2Yr Tuition*, there will be an increase of $1.216 in *Public 4yr Tuition*.

f)   The R^2 value is 44.29% (either calculated in a technology package or calculating the correlation r multiplied by itself). This means that 44.29% of the variation in the average tuitions of 4yr public colleges is accounted for by the regression on the average tuitions of 2yr public colleges.

53. Mutual fund flows.

a)   The y intercept is the value of the line when the x variable is zero. The intercept can be used as a starting point for predictions but it is not meaningful in all circumstances. In this equation, $\widehat{Flow} = 9747 - 771Return$ is of the form: $\hat{y} = b_0 + b_1 x$ where b_0 is the y intercept. The y intercept for this equation is 9747 ($M). This represents the value for money *Flow*ing into mutual funds when mutual fund performance *Return* is zero.

b)   The slope represents the change in y or the response variable for every x unit or one unit step in the predictor variable. The slope for this equation is 771, which means that for every 1 % increase in mutual fund *Return*, the *Flow* into mutual funds increases by 771 ($M).

c)   The predicted fund *Flow* for a month that had a market *Return* of 0% is the y intercept which has a value of 9747 ($M).

d)   If the recorded fund Flow was $5 billion during a month when the Return was 0%, the residual is calculated by subtracting the predicted value (calculated from the linear model equation) from the observed or measured value (*Residual = Data − Predicted* or $e = y - \hat{y}$). The predicted value of y at an x value of 0% was calculated in c) as 9747 ($M). $5 billion = 5000 $Million. The *Residual* = 5000 ($M) (*Data*) − 9747 ($M) (*Predicted*) = −4747 ($M). This model overestimated the *Flow* value.

55. Residual plots.

a)   The linear model seems appropriate. The residual plot has appropriate scatter of points and nothing remarkable.

b)   The linear model is not appropriate for this data set. The data points are curved indicating a nonlinear relationship.

c)   The linear model is not appropriate for this data set. The data points start out close together and then the spread increases as x increases.

57. Consumer spending. There are two influential outliers that give more weight to the linear regression (slope and intercept) and R^2 at 79%. The predictions will not be accurate for this regression. Looking at the scatterplot illustrates the reason why. Without these two data points, the R^2 drops to about 31%. The analyst should identify these two customers and examine why they are outliers. For analysis of the rest of the data points, these two customers should be set aside and the model refit to the rest of the data.

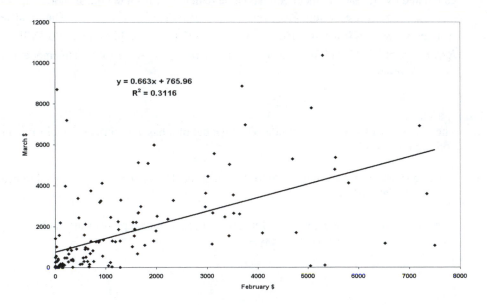

59. What slope? It is logical that the slope would reflect increased sales from an increase in advertising. It is also logical that it would be of the magnitude of increased sales of $10,000 from $1000 advertising. In the units of this problem, $10,000 is 0.01 $M. Therefore, the magnitude of the slope would be of the order of $0.01M ($y$ variable) per $1 thousand ($x$ variable) so 0.03 would be the closest in magnitude.

61. Misinterpretations.

   a) R^2 is an indication of the strength of a model but not the appropriateness of the model. The model may be influenced by other factors such as outliers or unusual data patterns. It is important to look at the scatterplot for these other factors.

   b) The statement should not be stated as an absolute fact. The annual sales figure is a prediction. The statement should be rephrased as, "The model predicts the quarterly sales will be $10M when $1.5 million is spent on advertising."

63. Business admissions. The four regression conditions are:

   a) Quantitative variable condition: Both variables are quantitative (*GPA* and *Starting Salary*).

   b) Linearity condition: Examine a scatterplot of *Starting Salary* (*y*) by *GPA* (*x*) to look for linearity.

   c) Outlier condition: Examine the scatterplot for outliers.

   d) Equal spread condition: This can be observed by plotting the regression residuals versus predicted values.

65. Used BMW prices.

   a)

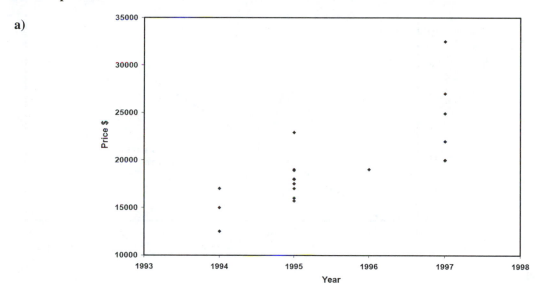

   b) There is a fairly strong positive association between used BMW 840 prices and their model year. There seems to be an increasing spread of data points as years increase. There is only one data point of year 1996.

   c) A linear model is appropriate for this relationship. It satisfies all requirements for a linear model. There is an increasing spread of data as years increase but there are not outliers or other unusual patterns.

   d) Correlation *r* which can be calculated by finding the square root of R^2. In the case, $\sqrt{.574} = 0.757627$ so r = 0.757.

   e) 57.4% of the variation in *Price* of a used BMW 849 can be accounted for by the *Year* the car was made.

f)   This relationship is not perfect. Other factors contribute to the variability of the Price, such as options, condition of car, and mileage.

67. Cost of living index.

a)   The association between the cost of living in 2007 and 2006 is linear, positive, and strong. The linearity of the scatterplot indicates that the linear model is appropriate.

b)   From R^2, 83.7% of the variability in cost of living in 2007 can be explained by the cost of living in 2006.

c)   The correlation is the square root of $R^2 = 0.837$. The correlation is $r = 0.915$.

d)   From the spreadsheet, we see that in 2006, Moscow had a cost of living of 123.9% of New York's in 2006, and from technology, we find the regression equation is:
Predicted 2007 index = 12.017 + 0.9425*(2006 Index). Thus for Moscow,
Predicted 2007 index for Moscow = 12.017 + 0.9425*123.9 = 128.793 or 128.8% of New York's.
The actual 2007 index was 134.4%. Thus the residual is 134.4% – 128.8% = about 5.6%.

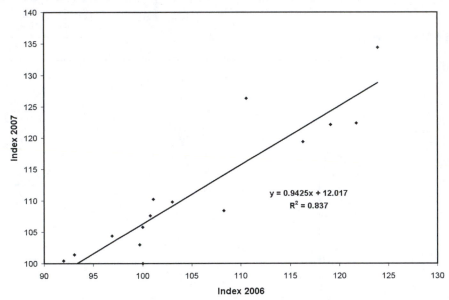

69. El Nino.

a)   The correlation r is the square root of R^2 (0.334) which equals 0.578.

b)   The meaning of R^2 in this context is that CO_2 levels account for 33.4% of the variation in *Mean Temperature*.

c)   The regression equation can be developed from the regression output:
$$\overline{MeanTemperature} = 15.3066 + 0.004\,CO_2.$$

d)   The slope of 0.004 means that the predicted *Mean Temperature* (in degrees Celsius) has been increasing at an average rate of 0.004 degrees (^{0}C)/ parts per million of CO_2.

e)   The intercept would be the value for the Mean Temperature would be 15.3066 if the ppm of CO_2 were zero but this doesn't make any sense.

f)   The residual scatterplot does not show anything remarkable, just a scattering about the zero point. There is no evidence of the violations of the assumptions of regression.

g) The *Mean Temperature* (^{0}C) prediction for 364 parts per million CO_2 levels is:

$$\overline{MeanTemperature} = 15.3066 + 0.004 * 364 = 16.7626\,^0C.$$

71. Dirt bikes.

    a) The scatterplot shows that the relationship between *Displacement* and *Total Weight* is not straight. Therefore, the Spearman's rank correlation would be more appropriate because it works with the ranks of data rather than their values.

    b) The relationship with the square root of *Displacement* is more nearly straight and would be more appropriate for fitting a linear model.

Chapter 7 – Randomness and Probability

SECTION EXERCISES

SECTION 7.1
1.

    a) Independent. There is no reason to think that the gender of one customer using an ATM machine will influence the gender of the next customer using an ATM machine (unless a large group of one gender comes to the ATM machine together).

    b) Independent. The last digit of one student's SS number provides no information about the last digit of another student's SS number.

    c) Not independent. Since all exams are taken by the same person, then knowing the performance on one exam provides information about the performance on subsequent exams.

SECTION 7.2
3.

    a) The strategy of always playing the numbers 1, 2, 3, 4, 5 will not improve the chances of winning the lottery (but it won't reduce the chances either). Since each number drawn is equally likely and independent of the others, this set of numbers is just as likely as any other set of numbers in the next drawing.

    b) The strategy of playing randomly generated numbers will not improve the chances of winning the lottery (but it won't reduce the chances either). Since each number drawn is equally likely and independent of the others, a set of randomly generated numbers is just as likely as any other set of numbers in the next drawing.

SECTION 7.4
5.

    a) P(car passes inspection) = 0.75

    b) Complement Rule:
P(car doesn't pass inspection) = 1 – P(car passes inspection) = 1 – 0.75 = 0.25

    c) Multiplication Rule:
P(both cars pass inspection) = 0.75×0.75 = 0.5625

    d) Addition Rule:
P(at least one car passes inspection) = 0.75+0.75-(0.5625) = 0.9375 or 1-(0.25^2) = 0.9375

SECTION 7.5
7.

    a) P(Serious Concern) = 760/1200 = 0.63

    b) This is a marginal probability.

    c) P(Republican *and* Serious Concern) = 210/1200 = 0.175

    d) This is a joint probability.

SECTION 7.6
9.

    a) Conditional probability:

$$P(\text{Serious Concern}|\text{Republican}) = \frac{P(\text{Serious Concern } and \text{ Republican})}{P(\text{Republican})} = \frac{210/1200}{500/1200} = 0.42$$

    b) Conditional probability:

$$P(\text{Republican}|\text{Serious Concern}) = \frac{P(\text{Serious Concern } and \text{ Republican})}{P(\text{Serious Concern})} = \frac{210/1200}{760/1200} = 0.2763$$

c)   Conditional probability:

$$P(\text{Serious Concern}|\text{Democrat}) = \frac{P(\text{Serious Concern and Democrat})}{P(\text{Democrat})} = \frac{440/1200}{500/1200} = 0.88$$

SECTION 7.7

11.

a)   By the Complement rule, 100% - 30% = 70%

b)   The 25% refers to the joint probability: P(Under 50 *and* Banks Online).

c)

Online Banking

Age	Yes	No	
Under 50	0.25	0.15	**0.40**
50 or Older	0.05	0.55	**0.60**
	0.30	**0.70**	**1.00**

d)   Conditional probability:

$$P(\text{Yes}|\text{Under 50}) = \frac{P(\text{Yes and Under 50})}{P(\text{Under 50})} = \frac{0.25}{0.40} = 0.625$$

e)   *Banking Online* and *Age* are not independent. The conditional probability of online banking for those under 50 is 0.625; this is not the same as the marginal probability of banking online (0.30). Consequently, knowing age changes the probability of banking online.

CHAPTER EXERCISES

13. What does it mean? part 1.

a)   Outcomes are equally likely and independent, that is, knowing one outcome will not affect the probability of the next. In simpler terms, on each spin, each number has the same chance of occurring as any other number. Additionally, there is an implication that the outcome is not determined through the use of an electronic number generator.

b)   This phrase is most likely a personal probability expressing his degree of belief that there will be a rate cut.

15. Airline safety.

a)   The overall probability of an airplane crash does not change due to recent crashes. The long term probability stays the same. Each airplane flight is independent of another flight.

b)   The overall probability of an airplane crash does not change due to a period in which there were no crashes. It makes no sense to say a crash is "due". If you say this, you are expecting probability to compensate for strange events in the past.

17. Fire insurance.

a)   It would be foolish to insure your neighbor's house for $300. Although you would probably collect the $300, there is a chance you could end up paying much more than $300. The risk is not worth the $300.

b)   The insurance company insures many people. The overwhelming majority of customers pays and never makes a claim. The few customers who do make a claim are offset by the many that simply send their premiums in regularly without a claim. The relative risk to the insurance company is low.

19. Toy company.
    a)   Yes (the sum of the probabilities = 1).

    b)   Yes (the sum of the probabilities = 1).

    c)   No (the sum of the probabilities > 1).

    d)   Yes (the sum of the probabilities = 1).

    e)   No (sum of the probabilities $\neq$ 1 and one value is negative).

21. Quality control. Assume that the defective tires are distributed randomly to all tire distributors so that the events can be considered independent. The multiplication rule may be used. P(at least one of four tires is defective) = $1 - P$(none are defective) = $1 - (0.98)(0.98)(0.98)(0.98) \approx 0.078$.

23. Auto warranty. All the events listed are disjoint. The addition rule can be used.
    a)   Subtract all repair probabilities from 1:
       P(no repairs) = $1 - P$(some repairs) = $1 - (0.17 + 0.07 + 0.04) = 1 - (0.28) = 0.72$

    b) P(no more than one repair) = P(no repairs or one repair) = $0.72 + 0.17 = 0.89$

    c) P(some repairs) = P(one or two or three or more repairs) = $0.17 + 0.07 + 0.04 = 0.28$

25. Auto warranty, part 2. Assuming that repairs on the two cars are independent from one another, the multiplication rule can be used. The probabilities from Exercise 11 are used in the calculations.
    a) P(neither will need repair) = $(0.72)(0.72) = 0.5184$

    b) P(both will need repair) = $(0.28)(0.28) = 0.0784$

    c) P(at least one will need repair) = $1 - P$(neither will need repair) = $1 - (0.72)(0.72) = 0.4816$

27. Auto warranty, again.
    a)   The repair needs for the two cars must be independent of each other.
    b)   This may not be reasonable. An owner may treat the two cars similarly, taking good (or poor) care of both. This may decrease (or increase) the likelihood that each needs to be repaired.

29. Real estate. The events are disjoint. Use the addition rule.
    a) P(pool or a garage) = P(garage) + P(pool) $- P$(both) = $0.64 + 0.21 - 0.17 = 0.68$

    b) P(neither a pool or garage) = $1 - P$(pool or garage) = $1 - 0.68 = 0.32$

    c) P(pool but no garage) = P(pool) $- P$(both) = $0.21 - 0.17 = 0.04$

31. Market research on energy.
    a) P("Increase Production") = $(342/1005) = 0.340$

    b) P("Equally Important" or "No Opinion") = $(50/1005) + (30/1005) = 80/1005 = 0.080$

33. Telemarketing contact rates. Assuming that the contacts are independent from one another, the multiplication rule can be used.
    a) P(contacted but refuse to cooperate) = P(contacted) $* P$(refuse to cooperate) = $(0.76)(0.62) = 0.4712$

    b) P(fail to contact or contact but refuse) = P(fail to contact) + P(contact but refuse) = $0.24 + (0.76)(0.62) = 0.7112$

c) P(fail to contact or contact but refuse) = 1 – P(interviewing) = 1 – (0.76)(0.38= 0.7112

35. Mars product information.
   a)   Assuming that all of the events are disjoint (an M&M cannot be two colors at once), use the multiplication rule where applicable.
      i. P(brown) = 1 – P(not brown) = 1 – P(yellow or red or orange or blue or green) = 1 – (0.20 + 0.20 + 0.10 + 0.10 +0.10) = 0.30

      ii. P (yellow or orange) = 0.20 + 0.10 = 0.30

      iii. P(not green) = 1 – P(green) = 1 – 0.10 = 0.90

      iv. P(striped) = 0

   b)   Since the events are independent (picking out one M&M doesn't affect the outcome of the next pick), the multiplication rule may be used.
      i. P(all three are brown) = (0.30)(0.30)(0.30) = 0.027

      ii. P(the third one is the first one that is red) = P(not red and not red and red) = (0.80)(0.80)(0.20) = 0.128

      iii. P(no yellow) = P(not yellow and not yellow and not yellow) = (0.80)(0.80)(0.80) = 0.512

      iv. P(at least one is green) = 1 – P(none are green) = 1 – (0.90)(0.90)(0.90) = 0.271

37. More Mars product information.
   a)   For one draw, the events of getting a red M&M and getting an orange M&M are disjoint events. Your single draw cannot be both red and orange.

   b)   For two draws, the events of getting a red M&M on the first draw and a red M&M on the second draw are independent events. Knowing that the first draw is red does not influence the probability of getting a red M&M on the second draw.

   c)   Disjoint events can be independent if one of the two events has a probability of 0 of occurring in the first place. For example, in the M&M example P(red and striped), if it is red it is not striped, they are disjoint. But knowing that it is not striped does not influence the P(red) and vice versa, since the event "striped" cannot occur – i.e. it has probability 0. Thus the events are also independent.

39. Tax accountant.
   a) P(all 3 audited) = (0.50)(0.50)(0.50) = 0.125

   b) P(none will be audited) = (0.50)(0.50)(0.50) = 0.125

   c) P(at least one will be audited) = 1 – (0.50)(0.50)(0.50) = 0.875

   d)   Assume that the events are independent of each other.

41. Information technology. Assuming that the mail delivery is independent from one day to the next, the multiplication rule may be used.
   a) P(gets interrupted on Monday and again on Tuesday) = (0.15)(0.15) = 0.0225

   b) P(gets interrupted for the first time on Thursday) = (0.85)(0.85)(0.85) (0.15) = 0.092

   c) P(gets stopped every day) = (0.15)(0.15)(0.15) (0.15)(0.15) = 0.00008

d) P(gets stopped at least once) = $1 - P$(never gets stopped) = $1 - (0.85)(0.85)(0.85)\,(0.85)(0.85) =$
   0.556

43. Casinos, part 2.
   a) Your thinking is correct. There are 47 cards left in the deck, 26 black and only 21 red.

   b) This is not an example of the Law of Large Numbers. The card draws are not independent of each other. The cards are not put back in the deck after they are drawn.

45. International food survey.
   a) P(person agreed) = (4228/7690) = 0.550

   b) P(person < 50 years old) = (1521+1525+1533+1513)/7690 = 0.792

   c) P(person < 50 years old *and* agrees) = (914+871+816+661)/7690 = 0.424

   d) P(< 50 years old *or* agrees) = P(<50 years old) + P(agrees) – P(<50 years old and agrees) = 0.792 + 0.550 - 0.424 = 0.918.

47. E-commerce. Conditional probabilities:
   a) $P(Male \mid Not\,Concerned) = \dfrac{P(Male\,and\,Not\,Concerned)}{P(Not\,Concerned)} = \dfrac{6}{18} = 0.333$

   b) $P(Not\,Concerned \mid Female) = \dfrac{P(Not\,Concerned\,and\,Female)}{P(Female)} = \dfrac{(18-6)}{(42-8-6)} = \dfrac{12}{28} = 0.429$

   c) $P(Female \mid Concerned) = \dfrac{P(Female\,and\,Concerned)}{P(Concerned)} = \dfrac{(24-8)}{(24)} = \dfrac{16}{24} = 0.667$

49. Pharmaceutical company.
   a) P(High BP and High Cholesterol) = 0.11

   b) P(High BP) = 0.11 + 0.16 = 0.27

   c) Conditional probability:
   $$P(High\,Chol \mid High\,BP) = \dfrac{P(High\,Chol\,and\,High\,BP)}{P(High\,BP)} = \dfrac{(0.11)}{(0.27)} = 0.407$$

   d) Conditional probability:
   $$P(High\,BP \mid High\,Chol) = \dfrac{P(High\,BP\,and\,High\,Chol)}{P(High\,Chol)} = \dfrac{(0.11)}{(0.32)} = 0.344$$

51. Pharmaceutical company, again. Does P(High BP|High Chol) = P(High BP)? From Exercise 37d), P(High BP|High Chol) = .344 and from the table P(High BP) = .11 + .16 = .27. P(High BP|High Chol) ≠ P(High BP); thus, they are not independent.

53. International food survey, part 2.
   a) P(13-19 years old who agrees) = (661/7690) = 0.086

   b) P(13-19 years old who agrees) = (661/1513) = 0.437

   c) P(agrees and is13-19 years old) = (661/4228) = 0.156

   d) P(disagrees and is 50+ years old) = (283/1627) = 0.174

  e) P(50+ years old and disagrees) = (283/1598) = 0.177

  f)   Check if P(Disagrees|> 50 years) = P(Disagrees)? From part e), P(Disagrees|> 50 years) = .177 and P(Disagrees) = 1627/7690 = .216.   Since P(Disagrees|> 50 years) $\neq$ P(Disagrees), they are not independent.

55. Real estate, part 2.

  a) P(Garage and No Pool) = P(Garage) – P(Both) = 0.64 - 0.17 = 0.47

  b)   Conditional probability: $P(Pool \mid Garage) = \dfrac{P(Garage \, and \, Pool)}{P(Garage)} = \dfrac{(0.17)}{(0.64)} = 0.266$

  c)   Check if P(Pool|Garage) = P(Pool). From part b) P(Pool|Garage) = 0.266 and from problem statement P(Pool) = 0.21. Since P(Pool|Garage) $\neq$ P(Pool), they are not independent.

  d)   Having a garage and a pool are not mutually exclusive events. P(Pool and Garage) = 0.17, not 0.

57. Telemarketing.

  a) P(customers reached by landline) = 1 – [P(customers with only cell) + P(customers with no service)] = 1 – [0.028 + 0.016] = 1 – 0.044 = 0.956 = 95.6%.

  b)   Check if P(Cell|Landline) = P(Cell). P(Cell) = 0.582 + 0.028 = 0.610.   P(Cell|Landline) = P(Cell and Landline)/P(Landline). P(Landline) = 1 - 0.016 - 0.028 = 0.956. Thus, P(Cell|Landline) = .582/.956 = .609. Since .609 $\neq$ .610, they are not independent although the probabilities are roughly the same.

59. Selling cars. Check if P(American|Student) = P(American). P(American|Student) = 107/195 = 0.549. P(American) = 212/359 = 0.591 Since 0.549 $\neq$ 0.591, they are not independent.

61. Used cars.

  a) P(first caller Jeep Liberty owner) = (23/149) = 0.154 or 15.4%

  b) P(first caller Jeep Liberty owner $18,000 – $18.999) = (17/149) = 0.114 or 11.4%

  c)   Conditional probability:

$$P(< \$19K \mid Jeep \, Liberty) = \frac{P(< \$19K \, and \, Jeep \, Liberty)}{P(Jeep \, Liberty)} = \frac{(17)}{(23)} = 0.739 = 73.9\%$$

  d)   Conditional probability:

$$P(Jeep \, Liberty \mid < \$19K) = \frac{P(Jeep \, Liberty \, and \, < \$19K)}{P(< \$19K)} = \frac{(17)}{(92)} = 0.185 = 18.5\%$$

Chapter 8 – Random Variables and Probability Models

SECTION EXERCISES

SECTION 8.1

1.

    a) Whether or not an employee includes a dependent child on his/her health insurance is a discrete variable.

    b) There are two possible values for this variable: Yes / No.

3. $E(X) = 0.2*0$ books $+ 0.4*1$ book $+ 0.4*2$ books $= 1.2$ books

SECTION 8.2

5.

$$SD(X) = \sqrt{(0-1.2)^2 * 0.2 + (1-1.2)^2 * 0.4 + (2-1.2)^2 * 0.4} = 0.748 books$$

7.

    a) $E(X) = 0.30*\$10 + 0.50*\$20 + 0.20*\$30 = \$19.$

    b) $SD(X) = \sqrt{(10-19)^2 \times 0.3 + (20-19)^2 \times 0.5 + (30-19)^2 \times 0.2} = \$7.0$

SECTION 8.3

9.

    a) $3X$: $\mu = 3*10 = 30$, $\sigma = 3* 2 = 6$

    b) $Y + 6$: $\mu = 20 + 6 = 26$, $\sigma = 5$

    c) $X + Y$: $\mu = 10 + 20 = 30$, $\sigma = \sqrt{2^2 + 5^2} = \sqrt{29} = 5.39$

    d) $X - Y$: $\mu = 10 - 20 = -10$, $\sigma = \sqrt{2^2 + 5^2} = \sqrt{29} = 5.39$

11.

    a) $E(X+10) = E(X) + 10 = \$100 + \$10 = \$110$; $SD(X + 10) = SD(X) = \$12$

    b) $E(5Y) = 5E(Y) = 5*\$90 = \$450$; $SD(5Y) = |5|SD(Y) = 5*\$8 = \$40$

    c) $E(X+Y) = E(X) + E(Y) = \$100 + \$90 = \$190$;
          To find $SD(X+Y)$ first find $VAR(X+Y)$
          $VAR(X+Y) = VAR(X) + VAR(Y) = 144$ $\$^2 + 64$ $\$^2 = 208$ $\$^2$
          $SD(X+Y) = \sqrt{208 \ \$^2} = \$14.42$

    d) In order to calculate the standard deviation (variance) in part (c), we must assume that the two different financial instruments (X and Y) must be independent.

SECTION 8.4

13. Yes, the Uniform model can be used because each of the outcomes has the same probability (1/6) of occurring.

15. Because we want to know how many calls it will take to find a firm in the U.S. owned by a woman (how many trials it will take to achieve the first success), the Geometric model should be used. The parameter is $p = 0.26$.

17.

    a) Find λ. An inspector examines 50 shirts, each with 6 buttons, in an hour. Therefore, 300 buttons are inspected per hour. Given that the probability of a button flaw is 0.002, then the mean number of button flaws per hour is $(0.002)(300) = 0.6 = \lambda$. Now find P(X=0) using the Poisson model:

$$P(X = 0) = \frac{e^{-0.6}0.6^0}{0!} = 0.5488$$

    b) Find $P(X \geq 1) = 1 - P(X = 0) = 1 - 0.5488 = 0.4512$

CHAPTER EXERCISES

19. New website.

    a) 1, 2, 3,...., n.

    b) Discrete (can list all outcomes).

21. Job interviews.

    a) 0, 1, 2, 3, 4

    b) Discrete (can list all outcomes).

    c) No, the outcomes are not equally likely.

23. Lottery.

    a) $\dfrac{1}{1000} \times \$350 + \dfrac{5}{1000} \times \$50 = \$0.35 + \$0.25 = \$0.60$

    b) $SD(X) = \sqrt{(350 - 0.6)^2 * 0.001 + (50 - 0.6)^2 * 0.005} = \$11.59$

    c) –$0.40

25. Commuting to work.

    a) Expected # of red lights = 0*0.05 + 1*0.25 + 2*0.35 + 3*0.15 + 4*0.15 + 5*0.05 = 2.25 lights.

    b) The standard deviation:

$$\sqrt{(2.25 - 0)^2 * 0.05 + (2.25 - 1)^2 * 0.25 + (2.25 - 2)^2 * 0.35 + (2.25 - 3)^2 * 0.15 + (2.25 - 4)^2 * 0.15 + (2.25 - 5)^2 * 0.05} = 1.26$$

27. Fishing tournament.

    a) No, the probability he wins the second changes depending on whether he won the first.

    b) The probability that he loses both tournaments: P(lose both) = P(lose 1st) * P(lose 2nd) = 0.6 * 0.7 = 0.42.

c)   The probability that he wins both tournaments:
P(win both) = P(win 1st) * P(win 2nd) = 0.4 * 0.2 = 0.08.

d)

X	0	1	2
P(X = x)	0.42	0.50	0.08

e)   E(X) = 0*0.42 + 1*0.50 + 2*0.08 = 0.66 tournaments;

$$SD(X) = \sqrt{(0-0.66)^2 *0.42 + (1-0.66)^2 *0.50 + (2-0.66)^2 *0.08} = 0.62 \text{ tournaments.}$$

29. Battery recall.

a)   No, because once you select a battery from your 10, the probability of the next battery being good or bad changes (shown in part b).

b)

Number good	0	1	2
P(number good)	$\left(\dfrac{3}{10}\right)\left(\dfrac{2}{9}\right) = \dfrac{6}{90}$	$\left(\dfrac{3}{10}\right)\left(\dfrac{7}{9}\right) + \left(\dfrac{7}{10}\right)\left(\dfrac{3}{9}\right) = \dfrac{42}{90}$	$\left(\dfrac{7}{10}\right)\left(\dfrac{6}{9}\right) = \dfrac{42}{90}$

c)   Expected # of good batteries: μ = E(X) = 0*(6/90) + 1*(42/90) + 2*(42/90) = 1.4 batteries.

d) $\sigma = \sqrt{(0-1.4)^2 *(6/90) + (1-1.4)^2 *(42/90) + (2-1.4)^2 *(42/90)} = 0.61$ batteries.

31. Commuting, part 2.

a)   Wait time: μ = E(X) = 5*(14.8) = 74.0 seconds.

b) $\sigma = \sqrt{9.2^2 + 9.2^2 + 9.2^2 + 9.2^2 + 9.2^2} \approx 20.57$ seconds. (Answers to standard deviation may vary slightly due to rounding of the standard deviation of the number of red lights each day.) The standard deviation may be calculated only if the days are independent of each other.

33. Insurance company.

a)   The standard deviation is large because the profits on insurance are highly variable. Although there will be many small gains, there will occasionally be large losses when the insurance company has to pay a claim.

b) μ = E(two policies) = 2*($150) = $300.
σ = SD(two policies) = $\sqrt{6000^2 + 6000^2}$ = $8485.28.

c) μ = E(1000 policies) = 1000*($150) = $150,000.
σ = SD(1000 policies) $\sqrt{1000*6000^2}$ ≈ $189,736.66.

d)   A natural disaster affecting many policyholders such as a large fire or hurricane.

35. Bike sale.

a)   B = # of basic models; D = # of deluxe models; Net Profit = $120B + $150D – $200.

b) μ = $120*5.4 + $150*3.2 – $200 = $928.

c) σ = SD(net profit) = $\sqrt{120^2 *1.2^2 + 150^2 *0.8^2}$ = $187.45.

d)   Mean – no; SD – yes (sales are independent).

37. Movie rentals.

   a) $\mu = 2*1.3 = 2.6$ days.

   $\sigma = $ SD(round trip time) $= \sqrt{0.5^2 + 0.5^2} = 0.707$ days.

   b) $\mu = 2*1.3 + 1.1 = 3.7$ days.

   $\sigma = $ SD(combined time) $= \sqrt{0.5^2 + 0.5^2 + 0.3^2} = 0.768$ days.

   c) The probability that time > 9 days to complete the total rent cycle: Compare 9 days to the time cycle from b) of 3.7 days added to the 3.7 days that the customer holds the DVD:

   $$z = \frac{x - \mu}{\sigma} = \frac{(9 - (3.7 + 3.7))}{\sqrt{2^2 + 0.768^2}} = 0.7468 \text{ for a P-value} = 0.2276 \ (0.2266 \text{ from tables}).$$

39. eBay.

   a) Let $X_i = $ price of ith Hulk figure sold; $Y_i = $ price of ith Iron Man figure sold; Insertion Fee = $0.55;
   T = Closing Fee = 0.875 $(X_1 + X_2 + \ldots + X_{19} + Y_1 + \ldots + Y_{13})$
   Net Income = $(X_1 + X_2 + \ldots + X_{19} + Y_1 + \ldots + Y_{13}) - 32(0.55) - 0.0875 \ (X_1 + X_2 + \ldots + X_{19} + Y_1 + \ldots + Y_{13})$.

   b) $\mu = $ E(net income) = (19*\$12.11 + 13*\$10.19) $- 32*\$0.55 - 0.0875*$Total Sales = \$313.24
   where Total Sales = 19*\$12.11 + 13*\$10.19 = \$362.56

   c) SD(net income): $\sigma = \sqrt{19*1.38^2 + 13*0.77^2} = \$6.625$

   d) Yes, to compute the standard deviation.

41. Bernoulli.

   a) No, these are not Bernoulli trials. The possible outcomes are 1, 2, 3, 4, 5, and 6. There are more than two possible outcomes.

   b) Yes, these may be considered Bernoulli trials. There are only two possible outcomes: Type A and not Type A. Assuming that the 120 donors are representative of the population, the probability of having Type A blood is 43%. The trials are not independent because the population is finite, but the 120 donors represent less than 10% of all possible donors.

   c) No, these are not Bernoulli trials. The probability of choosing a man changes after each promotion and the 10% condition is violated.

   d) No, these are not Bernoulli trials. We are sampling without replacement, so the trials are not independent. Samples without replacement may be considered Bernoulli trials if the sample size is less than 10% of the population, but 500 is more than 10% of 3000.

   e) Yes, these may be considered Bernoulli trials. There are only two possible outcomes: sealed properly and not sealed properly. The probability that a package is unsealed is constant at about 10%, as long as the packages checked are a representative sample of all packages. In addition, the 24 packages are less than 10% of population.

43. Closing sales.

   a) Probability that he fails to close for the 1st time on his 5th attempt $= 0.80^4 * 0.20 = 0.0819$.

   b) Probability that he closes his 1st presentation on his 4th attempt $= 0.20^3 * 0.80 = 0.0064$.

   c) Probability that the 1st presentation he closes is on his 2nd attempt $= 0.20 * 0.80 = 0.16$.

   d) Probability that the 1st presentation he closes is on one of his first 3 attempts is calculated by using the probabilities for the possible configurations for the 3 attempts: P(1) + P(2) + P(3) = 0.8 + (0.2)(0.8) + (0.2)(0.2)(0.8) = 0.992.

45. Side effects. E(X): (1/p) = 1/0.07 which means that we expect 14.28 or about 15 patients.

47. Missing pixels.

   a) The mean number of pixels per square foot = 4.7 blank pixels/60 sq ft = 0.0783 pixels per sq ft.

   b) The standard deviation of pixels per square foot $\sqrt{0.0783} = 0.280$ pixels.

   c) Probability that a 2ft by 3ft screen has at least one defect: 0.0783 pixels/sq ft * 6 = 0.47 defects. Add probabilities using the Poisson model for X = 1, 2, 3, +...

$$P(X=1) = \frac{e^{-0.47}0.47^1}{1!} = 0.294 \;;\; P(X=2) = \frac{e^{-0.47}0.47^2}{2!} = 0.069 \;;\; P(X=3) = 0.0108;$$

$$P(X=4) = 0.0013; +\ldots = 0.375.$$

   d) Probability that a 2ft by 3ft screen will be replaced (> 2 pixels): $P(X>2) = \frac{e^{-0.47}0.47^3}{3!} = 0.0108$

       Add probabilities for X = 4, 5, +...= 0.012.

49. Hurricane insurance.

   a) The probability of having a year in Florida with no hurricane hits:

       (22 storms/17 yr) = 1.294 storms/yr: $P(X=0) = \dfrac{e^{-1.294}1.294^0}{0!} = 0.2742$

   b) The probability of exactly one hit: $P(X=1) = \dfrac{e^{-1.294}1.294^1}{1!} = 0.3548$

   c) The probability of more than 3 hits: $P(X>3) = \dfrac{e^{-1.294}1.294^4}{4!} + P(X=5) + P(X=6), etc. = 0.0425$

51. Professional tennis.

   a) The probability that all 6 first serves will be in: $0.67^6 = 0.0905$.

   b) The probability that exactly 4 first serves will be in – use binomial model:

$$\frac{n!}{x!(n-x)}p^x q^{n-x} = \frac{6!}{4!(6-4)} *0.67^4 *0.33^2 = 0.329$$

   c) P(exactly 4 successes in 6 trials) (use binomial formula): $\dfrac{n!}{x!(n-x)} = \dfrac{6!}{4!(6-4)} = 0.687.$

53. Satisfaction survey.

   a) A uniform distribution; all numbers should be equally likely to be selected.

   b) The probability that the number selected will be an even number: 0.50 or 50%.

   c) The probability that the number selected will end in 000: 10/9999 = 0.001.

55. Web visitors.

   a) The Poisson model because it is a good model to use when the data consists of counts of occurrences. The events must be independent and the mean number of occurrences stays constant.

   b) Probability that in any one minute at least one purchase is made:

$$P(X=1) + P(X=2) +\ldots = \frac{e^{-3}3^1}{1!} + \frac{e^{-3}3^2}{2!} +\ldots = 0.9502$$

   c) Probability that no one makes a purchase in the next 2 minutes:

$$P(X=0) = \frac{e^{-6}6^0}{0!} = 0.0025$$

Chapter 9 – The Normal Distribution

SECTION EXERCISES

SECTION 9.1

1. In order to determine the placement exam upon which an incoming MBA student did better relative to other students, convert each grade to a z – score by subtracting out the mean and dividing by the standard deviation as follow:

   For the economics placement exam, we have

   $$z = \frac{82 - 72}{8} = 1.25$$

   For the mathematics placement exam, we have

   $$z = \frac{86 - 68}{12} = 1.5$$

   Therefore, the MBA student scored 1.25 standard deviations above the mean on the economics placement exam and .1.50 standard deviations above the mean on the mathematics placement exam; she did "better" on the mathematics exam.

3. A z-score of 2.20 means that you scored 2.2 standard deviations above the average on the Executive Aptitude test.

5.
   a) According to the 68-95-99.7 rule, only 5% of the distribution is beyond 2 standard deviations away from the mean. Consequently, only 2.5% of the distribution is 2 standard deviations above the mean. Therefore, less than 2.5% of the distribution is above a z-score of 2.2. Since the cutoff to qualify for admission to the executive training program is the top 3%, a z-score of 2.2 qualifies.
   b) In order to apply the 68-95-99.7 rule, we must assume that the distribution of test scores is unimodal and symmetric.

SECTION 9.2

7.

   a)

b)   We would expect the central 68% of automobiles to be within the interval of 18.6 to 31.0 mpg.

c)   About 16% of automobiles should get more than 31 mpg. Since we know that 32% of the automobiles will be outside the central 68% interval (18.6 to 31.0 mpg.) and that the distribution is symmetric, then ½ of 32% or 16% will be above 31 mpg.

d)   About 13.5% of the automobiles should get between 31 and 37.2 mpg. From above we can see that the percentage falling in this range is ½ (95% - 68%) = ½ (27%) = 13.5%.

e)   The gas mileage of the worst 2.5% of all automobiles is less than 12.4 mpg.

9.

a) $P(z > 1.5) = 1 - 0.9332 = .0668$ or 6.68%

b) $P(z < 2.25) = 0.9878$ or 98.78%

c) $P(-1 < z < 1.15) = P(z < 1.15) - P(z < -1) = 0.8749 - 0.1587 = 0.7162$ or 71.62%

d) $P(|z| > 0.5) = P(z < -0.5) + P(z > 0.5) = 0.3085 + (1 - 0.6915) = 0.617$ or 61.7%

11.

a)   Find ? such that $P(z > ?) = 0.20$ which is the same as $P(z < ?) = 0.80$.
The closest probability in Table Z is 0.7995 which corresponds to $z = 0.84$ (or 0.842 with interpolation).   Therefore, the highest 20% falls above $z = 0.84$.

b)   Find ? such that $P(z > ?) = 0.75$ which is the same as $P(z < ?) = 0.25$.
The closest probability in Table Z is 0.2514 which corresponds to $z = -0.67$ (or -0.675 with interpolation). Therefore the highest 75% falls above $z = -0.67$.

c)   Find ? such that $P(z < ?) = 0.03$.
The closest probability in Table Z is 0.0301 which corresponds to $z = -1.88$ (or -1.881 with interpolation).   Therefore the lowest 3% falls below $z = -1.88$.

d)   Find ? such that $P(? < z < ?) = 0.90$.
If the probability in the middle is 0.90, then there is 0.05 left in each tail. Therefore, find $P(z < ?) = 0.05$ and $P(z < ?) = 0.95$. The probability 0.05 falls exactly between 0.0495 and 0.0505 which corresponds to $z = -1.645$; and the probability 0.95 falls exactly between 0.9495 and 0.9505 which corresponds to $z = 1.645$. Therefore the middle 90% falls between -1.645 and 1.645.

SECTION 9.3

13. Yes, a Normal model appears appropriate for the speeds of cars. The histogram of the data shows a unimodal and symmetric distribution and the Normal probability plot is straight.

SECTION 9.4

15.

a) $\mu = 164$ miles as the expected value of the miles remaining is the same for the new set of tires.

$$\sigma = \sqrt{14^2 + 14^2} = \sqrt{392 miles^2} \approx 19.799 \text{ miles}$$

b) Find the probability that the set of tires will last another 164 miles (till the end of the 500 mile race). Since tire mileage is normal, and we know the expected value and standard deviation of the mileage remaining after two changes, we can convert 164 to its corresponding z-score as follows:

$$z = \frac{164 - 168}{19.799} = -0.20; \text{ find } P(z > -0.20) = 1 - P(z < -0.20) = 1 - 0.4207 = 0.5793.$$

SECTION 9.5

17.

a) Probability of at least 246 passengers showing up can be evaluated using the Normal approximation assuming the conditions are satisfied:

$$z = \frac{x - \mu}{\sigma} = \frac{246 - (255 - 12.75)}{3.48} = 1.077 \text{ using } \mu = np = 255 \times 0.05 = 12.75$$

$$\text{and } \sigma = \sqrt{np(1-p)} = \sqrt{255 \times 0.05(1 - 0.05)} = 3.48$$

$P(z > 1.077) = 0.140$ (0.175 with continuity correction).

b) Answers may vary. That's a fairly high proportion, but the decision depends on the relative costs of not selling seats and bumping passengers.

SECTION 9.6

19.

a) A uniform distribution; all numbers should be equally likely to be selected.

b) The probability that the number selected will be for an incubator business: 200/10,000 = 0.02 or 2%.

c) The probability that the number selected will be above 9999: 999/10,000=.0999 or 9.99%.

CHAPTER EXERCISES

21. Mutual fund returns.

a) Returns of 8.0% or more = one standard deviation: 2.4% (mean) + 5.6% (value of the standard deviation). Using the 68-95-99.7 rule, +/- one standard deviation accounts for the middle 68% of the data, leaving 32% divided by 2 (tails on both sides). Therefore, 16% of the returns are 8.0% or more (area to the right of one standard deviation).

b) Returns of 2.4% or more = mean value: 2.4% (mean). The mean is in the center of the distribution; therefore, 50% of the returns are 2.4% or more (area to the right of the mean).

c) Returns between -8.8% and 13.6% = two standard deviations: 2.4% (mean) + 2(5.6%) = 13.6%; 2.4% (mean) – 2(5.6%) = -8.8%. Using the 68-95-99.7 rule, +/- two standard deviations account for the middle 95% of the data.

d) Returns of 19.2% or more = three standard deviations: 2.4% (mean) + 3(5.6%) = 19.2%. Using the 68-95-99.7 rule, +/- three standard deviations accounts for the middle 99.7% of the data, leaving 0.30% divided by 2 (tails on both sides). Therefore, 0.15% of the returns are 19.2% or more (area to the right of three standard deviations).

23. Mutual funds, again.
    a)   The cutoff return value that would separate the highest 50% is the area to the right of the mean of 2.4% (center of the distribution).

    b)   The cutoff return value that would separate the highest 16% is the area to the right of one standard deviation to the right of the mean, which = 8.0%.

    c)   The cutoff return value that would separate the lowest 2.5% is the area to the left of two standard deviations to the left of the mean, which = -8.8%.

    d)   The middle 68% is the area between one standard deviation to the left of the mean (2.4% – 5.6% = -3.2%) and one standard deviation to the right of the mean (2.4% + 5.6% = 8.0%). Therefore, the cutoff return values that would separate the middle 68% = (-3.2% < x < 8.0%).

25. Currency exchange rates.
    a)   The probability of x < 1.459 euros is equivalent to the area to the left of the mean which = 50%.

    b)   The probability of x > 1.492 euros is equivalent to the area to the right of one standard deviation to the right of the mean (1.459 + 0.033) which = 16%.

    c)   The probability of x < 1.393 euros is equivalent to the area to the left of two standard deviations to the left of the mean (1.459 – 2(0.033)) which = 2.5%.

    d)   The probability of x < 1.410 euros is between one and two standard deviations to the left of the mean (between 1.393 and 1.426). The probability of x > 1.542 euros is between two and three standard deviations to the right of the mean (between 1.525 and 1.558). It is more unusual to have a day on which the pound was worth more than 1.542 (further away from the mean).

27. Currency exchange rates, again.
    a)   The highest 16% of EUR/GBP rates is equivalent to the area to the right of the value of one standard deviation to the right of the mean = 1.459 + 0.033 = 1.492 or x > 1.492.

    b)   The lowest 50% of the rates are to the left of the mean or x < 1.459.

    c)   The middle 95% of the rates are between the values for two standard deviations to the left and to the right of the mean or 1.393 < x < 1.525.

    d)   The lowest 2.5% of the rates are two standard deviations to the left of the mean or 1.459 – 2(0.033) = 1.393 or x < 1.393.

29. Mutual fund probabilities. Use technology to solve.
    a)   The percent of funds > 6.8% (0.068) = 21.6%.

    b)   The percent of funds between 0% and 7.6% = 48.9%.

    c)   The percent of funds > 1% (0.01) = 59.9%.

    d)   The percent of funds < 0% = 33.4%.

31. Mutual funds, once more. Use technology to solve.
    a)   The cutoff value x for the highest 10% = 9.57% or x > 9.58%.

    b)   The cutoff value x for the lowest 20% = -2.31% or x < -2.31%.

c)   The cutoff values for the middle 40% = (-0.54% < *x* < 5.34%).

d)   The cutoff value *x* for the highest 80% = -2.31% or *x* > -2.31%.

33. Mutual funds, finis. Use technology to solve.
   a)   The 40th percentile is represented by the lowest 40% (or upper 60%) = 0.98%.

   b)   The 99th percentile is represented by the lowest 99% (or upper 1%) = 15.4%.

   c)   The IQR represents distribution between Q1 (25th percentile) and Q3 (75th percentile) = (-1.38 % < *x* < 6.18%). The area = 6.18% - (-1.38%) = 7.56%.

35. Parameters.

a)

$$z = \frac{y - \mu}{\sigma}$$

$$0.126 = \frac{30 - 20}{\sigma}$$

$$0.126\sigma = 10$$

$$\sigma = 79.58$$

b)

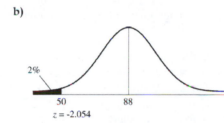

$$z = \frac{y - \mu}{\sigma}$$

$$-2.054 = \frac{50 - 88}{\sigma}$$

$$-2.054\sigma = -38$$

$$\sigma = 18.50$$

c)

$$z = \frac{y - \mu}{\sigma}$$

$$0.842 = \frac{100 - \mu}{5}$$

$$(0.842)(5) = 100 - \mu$$

$$\mu = 95.79$$

d)

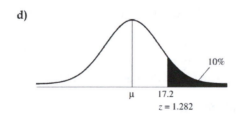

$$z = \frac{y - \mu}{\sigma}$$

$$1.282 = \frac{17.2 - \mu}{15.6}$$

$$(1.282)(15.6) = 17.2 - \mu$$

$$\mu = -2.79$$

37. SAT or ACT? The applicant scoring 650 on the SAT: $z = \dfrac{(y-\mu)}{\sigma} = \dfrac{(650-550)}{75} = 1.33$. The applicant

scoring 33 on the ACT: $z = \dfrac{(y-\mu)}{\sigma} = \dfrac{(33-27)}{3} = 2.0$. The ACT score is a better score because the z-value

is larger, indicating that the score is farther above the mean in standard deviation units than the SAT score.

39. Claims.

    a) You would like to know about the standard deviation to know about the battery consistency and how long they might last. Standard deviation measure variability, which translates to consistency in everyday use. A type of battery with a small standard deviation would be more likely to have life spans close to their mean life span than a type of battery with a larger standard deviation.

    b) The second company's batteries have a higher mean life span, but a larger standard deviation, so they have more variability. The decision is not clear-cut. The z-value for lasting less than 21 months is $\dfrac{(21-24)}{1.5} = -2.0$ with probability $< 2.5\%$ for the first company and $\dfrac{(21-30)}{9} = -1.0$ with probability $< 16\%$ for the second company. The chances of lasting more than 3 years are zero for the first company (z-value $= \dfrac{(36-24)}{1.5} = 8.0$) and $> 16\%$ for the second company (z-value $= \dfrac{(36-30)}{9} = 0.67$). For longevity, the edge goes to the second company.

41. CEOs. The standard deviation of the distribution of years of industry experience for CEOs must be 6 years. CEOs can have between 0 and 40 (or possibly 50) years of experience. A workable standard deviation would cover most of that range of values with +/- 3 standard deviations around the mean. If the standard deviation were 6 months, some CEOs would have years of experience 10 or 20 standard deviations away from the mean, whatever it is. That isn't possible. If the standard deviation were 16 years, +/- 2 standard deviations would be a range of 64 years. That's way too high. The only reasonable choice is a standard deviation of 6 years in the distribution of years of experience.

43. Web visitors.

    a) The Poisson model, if conditions are satisfied, would be appropriate to model the number of purchases per minute.

    b) If we can assume that the Poisson applies to model the number of purchases per minute, then the exponential model should be used for the time between purchases.

    c) $1/\lambda$ or 1/3 minutes.

    d) Use the exponential distribution with $\lambda = 3$ to find $P(1 < X < 2)$.

$$P(1 < x < 2) = e^{-(3)(1)} - e^{-(3)(2)} = 0.04978 - 0.00248 = 0.0473$$

45. Lefties.

Find the probability that there will be more than 170 righties in a typical class.

$$z = \frac{x-\mu}{\sigma} = \frac{170 - (188 - 24.44)}{4.611} = 1.40 \text{ using } \mu = np = 188 \times 0.13 = 24.44$$

and $\sigma = \sqrt{np(1-p)} = \sqrt{188 \times 0.13(1-0.13)} = 4.611$

$P(z > 1.40) = 0.0808$

47. Rickets.

$$\text{Find } P(x \le 50) = P\!\left(z \le \frac{50 - (320)(0.20)}{\sqrt{(320)(0.20)(0.80)}}\right) = P(z \le -1.96) \approx 0.975$$

49. Low job satisfaction. Any job satisfaction score more than two standard deviations below the mean (less than $100 - 2(12) = 76$) might be considered unusually low. We would expect to find someone with a job satisfaction score less than three standard deviations (less than $100 - 3(12) = 64$) very rarely.

51. Management survey.
    a)   About 16% of managers will exercise fewer than one standard deviation below the mean number of hours.

    b)   Using these data, one standard deviation below the mean is $3.66 - 4.93 = -1.27$ hours, which is impossible.

    c)   Because the standard deviation is larger than the mean, the distribution is strongly skewed to the right and not symmetric.

53. Drug company.
    a)   The Normal model:

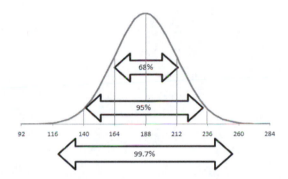

    b)   From technology, the percent of adult women expected to have cholesterol levels over 200 mg/dL is 30.85%.

    c)   The percent of adult women expected to have cholesterol levels between 150 and 170 mg/dL is 17.00%.

    d)   The interquartile range (IQR = Q3 – Q1) of cholesterol levels is calculated as (204.19 to 171.81) = 32.38.

    e)   The highest 15% of women's cholesterol levels are above 212.87 points.

Chapter 10 – Sampling Distributions

SECTION 10.1

1.

    a) The sample size is large enough to expect that the shape of the sampling distribution is still normal and the mean of the sampling distribution would be the same.

    b) The mean of the sampling distribution is 0.36 (the true proportion of smart phone users).

    c) Increasing the sample size from 200 to 500 does not change the answers in parts a) and b). The shape and mean of the sampling distribution would be the same.

SECTION 10.2

3.

    a) $SD(\hat{p}) = \sqrt{\dfrac{p(1-p)}{n}} = \sqrt{\dfrac{0.36(1-0.64)}{200}} = 0.0339$

    b) $P(\hat{p} > 0.36) = P(z > \dfrac{0.36 - 0.36}{0.0339} = 0) = P(z > 0) = 0.5$

    c) $P(0.30 < \hat{p} < 0.40) = P(\dfrac{0.30 - 0.36}{0.0339} < z < \dfrac{0.40 - 0.36}{0.0339}) = P(-1.77 < z < 1.18) = 0.8810 - 0.0392 = 0.842$

    d) $P(\hat{p} < 0.28) = P(z < \dfrac{0.28 - 0.36}{0.0339}) = P(z < -2.36) = 0.0091$

    e) $P(\hat{p} > 0.42) = P(z > \dfrac{0.42 - 0.36}{0.0339}) = P(z > 1.77) = 1 - 0.9616 = 0.0384$

5. Yes. Assuming that the survey is random, the homeowners should be independent. Since we do not know the true proportion, we cannot check the success/failure condition. However, we have observed 10 successes which is sufficient.

SECTION 10.3

7. The mean of the sampling model is $3.48 with a standard deviation of $\dfrac{\$2.23}{\sqrt{20}} = \$0.50$.

SECTION 10.4

9.

    a) The sampling distribution of the mean would be normal. The sample size does not matter here because cholesterol levels are roughly normally distributed.

    b) The mean of the sampling distribution is 215 mg/dL.

    c) $SD(\bar{y}) = \dfrac{\sigma}{\sqrt{n}} = \dfrac{30mg/dL}{\sqrt{42}} = 4.63mg/dL$

    d) If the sample size is increased to 100, only the answer to c) would change.

$$SD(\bar{y}) = \frac{\sigma}{\sqrt{n}} = \frac{30mg/dL}{\sqrt{100}} = 3.0mg/dL$$

SECTION 10.5

11.

    a) $SE(\hat{p}) = \sqrt{\dfrac{\hat{p}\hat{q}}{n}} = \sqrt{\dfrac{(0.15)(0.85)}{100}} = 0.0357$

    b) She would need a sample size of 400. In this way she would have $\sqrt{400} = 20$ in the denominator (twice the current denominator) which would cut the standard deviation by half.

13.

    a) A skewed to the right model makes sense here because we would expect many small fish and only a few very large ones to inhabit the lake.

    b) We do not know the exact distribution, but we do know that it is not normal because it is skewed to the right. Since we can't assume normality, and we don't know the exact distribution, we cannot determine the probability of a randomly selected (caught) bass weighing over 3 pounds.

    c) We know that the distribution of weights is skewed to the right. A sample size of 5 is not large enough to use the Central Limit Theorem for the sampling distribution of the mean. Without being able to assume a normal sampling distribution, we cannot determine the probability that the mean weight of someone's catch is over 3 pounds.

    d) A sample size of 60 allows us to use the Central Limit Theorem. Therefore the sampling

    distribution of the mean is normal with a standard deviation of $SD(\bar{y}) = \dfrac{\sigma}{\sqrt{n}} = \dfrac{23.2\,pounds}{\sqrt{60}} = 2.995$

    or 3 pounds.

    e) No, this would not be surprising. A sample mean weight of 4.5 pounds is only about 1/3 standard deviations above the mean of the sampling distribution. There is about a 37% that the sample mean weight would be more than 4.5 pounds. The 68-95-99.7 Rule does not apply.

CHAPTER EXERCISES

15. Send money.

All of the histograms are centered around $p = 0.05$. As n gets larger, the shape of the histograms get more unimodal and symmetric, approaching a Normal model, while the variability in the sample proportions decreases.

17. Send money, again.

a)

n	Observed mean	Theoretical mean	Observed st. dev.	Theoretical standard deviation
20	0.0497	0.05	0.0479	$\sqrt{(0.05)(0.95)/20} \approx 0.0487$
50	0.0516	0.05	0.0309	$\sqrt{(0.05)(0.95)/50} \approx 0.0308$
100	0.0497	0.05	0.0215	$\sqrt{(0.05)(0.95)/100} \approx 0.0218$
200	0.0501	0.05	0.0152	$\sqrt{(0.05)(0.95)/200} \approx 0.0154$

b)   All of the values seem very close to what we would expect from theory.

c)   The histogram for $n = 200$ looks quite unimodal and symmetric. We should be able to use the Normal model.

d)   The success/failure condition requires np and nq to both be at least 10, which is not satisfied until $n = 200$ for $p = 0.05$. The theory supports the choice in part c.

19. Stock picking.

a)   The histogram of these proportions is expected to be symmetric, but **not** because of the Central Limit Theorem. The sample of 16 dart throws is not large. The distribution of these proportions is expected to be symmetric because the probability that the stock went up is the same as the probability that the stock went down.

b)   The histogram is expected to have its center at 0.5, the probability that the stock went up.

c)   The standard deviation of data displayed in this histogram should be approximately equal to the standard deviation of the sampling distribution model, $\sqrt{\dfrac{pq}{n}} = \sqrt{\dfrac{(0.5)(0.5)}{16}} = 0.125$.

d)   The Success/Failure condition is not met. The Normal model is not appropriate in this case.

21. Bigger portfolio.

a) $\mu_{\hat{p}} = p = 0.5$ and $\sigma(\hat{p}) = \sqrt{\dfrac{pq}{n}} = \sqrt{\dfrac{(0.5)(0.5)}{25}} = 0.1$

About 68% of the sample proportions are expected to be between 0.4 and 0.6, about 95% are expected to be between 0.3 and 0.7, and about 99.7% are expected to be between 0.2 and 0.8.

b)   First of all, the generated random numbers used to pick the stocks are independent of one another. There is no need to check the 10% Condition. Second, $np = nq = 12.5$, so both are greater than 10. The Success/Failure condition is met, so the sampling distribution model is $N(0.5, 0.1)$.

c) $\mu_{\hat{p}} = p = 0.5$ and $\sigma(\hat{p}) = \sqrt{\dfrac{pq}{n}} = \sqrt{\dfrac{(0.5)(0.5)}{64}} = 0.0625$

About 68% of the sample proportions are expected to be between 0.4375 and 0.5625, about 95% are expected to be between 0.375 and 0.625, and about 99.7% are expected to be between 0.3125 and 0.6875.

The stock picks are independent of one another, and $np = nq = 32$, so both are greater than 10. The Success/Failure condition is met, so the sampling distribution model is $N(0.5, 0.0625)$.

d)   As the number of stock picks increases, the sampling distribution model will still be Normal and centered at 0.5, but the standard deviation will decrease. The sampling distribution model will be less spread out.

23. A winning investment strategy?

a) For 200 dart throws, the sampling distribution model is Normal with $\mu_{\hat{p}} = p = 0.5$ and $\sigma(\hat{p})$
$= \sqrt{\dfrac{pq}{n}} = \sqrt{\dfrac{(0.5)(0.5)}{200}} \approx 0.0354$. His sample proportion of $\hat{p} = 0.58$ is about 2.26 standard deviations above the expected proportion, which is unusual, but not extraordinary. According to the Normal model, we expect sample proportions this high or higher about 1.2% of the time.

b) The expected number of students out of 100 to be this successful is 1.2. So it not surprising that one out of 100 would be this successful.

25. Speeding.

a) $\mu_{\hat{p}} = p = 0.70$

$\sigma(\hat{p}) = \sqrt{\dfrac{pq}{n}} = \sqrt{\dfrac{(0.7)(0.3)}{80}} \approx 0.051$.

About 68% of the sample proportions are expected to be between 0.649 and 0.751, about 95% are expected to be between 0.598 and 0.802, and about 99.7% are expected to be between 0.547 and 0.853.

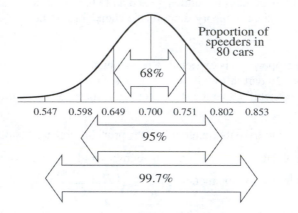

b) **Randomization condition:** The sample may not be representative. If the flow of traffic is very fast, the speed of the other cars around may have some effect on the speed of each driver. Likewise, if traffic is slow, the police may find a smaller proportion of speeders than they expect.

10% condition: 80 cars represent less than 10% of all cars

Success/Failure condition: $np = 56$ and $nq = 24$ are both greater than 10.

The Normal model may not be appropriate. Use caution. (And don't speed!)

27. Vision.

a) **Randomization condition:** Assume that the 170 children are a representative sample of all children.

10% condition: A sample of this size is less than 10% of all children.

Success/Failure condition: $np = 20.4$ and $nq = 149.6$ are both greater than 10.

Therefore, the sampling distribution model for the proportion of 170 children who are nearsighted is $N(0.12, 0.025)$.

b) The Normal model is to the right.

c) They might expect that the proportion of nearsighted students to be within 2 standard deviations of the mean.

According to the Normal model, this means they might expect between 7% and 17% of the students to be nearsighted, or between about 12 and 29 students.

29. Loans.

a) $\mu_{\hat{p}} = p = 7\%$

$\sigma(\hat{p}) = \sqrt{\dfrac{pq}{n}} = \sqrt{\dfrac{(0.07)(0.93)}{200}} \approx 1.8\%$

b) **Randomization condition:** Assume that the 200 people are a representative sample of all loan recipients.

10% condition: A sample of this size is less than 10% of all loan recipients.

Success/Failure condition: $np = 14$ and $nq = 186$ are both greater than 10.

Therefore, the sampling distribution model for the proportion of 200 loan recipients who will not make payments on time is $N(0.07, 0.018)$.

c) According to the Normal model, the probability that over 10% of these clients will not make timely payments is approximately 0.048.

$$z = \frac{\hat{p} - \mu_{\hat{p}}}{\sqrt{\dfrac{pq}{n}}}$$

$$z = \frac{0.10 - 0.07}{\sqrt{\dfrac{(0.07)(0.93)}{200}}}$$

$$z \approx 1.663$$

31. Back to school?

Randomization condition: We are considering colleges with freshman classes of 400 students. These are not random samples, and not all of the colleges considered may be typical of all colleges. We should be careful using this sampling distribution model.

10% Condition: 400 students is less than 10% of all college students.

Success/Failure condition: $np = 296$ and $nq = 104$ are both greater than 10.

Therefore, the sampling distribution model for $\hat{p}$ is Normal, with:

$\mu_{\hat{p}} = p = 0.74$

$\sigma(\hat{p}) = \sqrt{\dfrac{pq}{n}} = \sqrt{\dfrac{(0.74)(0.26)}{400}} \approx 0.022$

According to the sampling distribution model, about 68% of the colleges are expected to have retention rates between 0.718 and 0.762, about 95% of the colleges are expected to have retention rates between 0.696 and 0.784, and about 99.7% of the colleges are expected to have retention rates between 0.674 and 0.806. However, the conditions for the use of this model may not be met. We should be cautious about making any conclusions based on this model.

33. Back to school, again.

Provided that the students at this college are typical, the sampling distribution model for the retention rate, $\hat{p}$, is Normal with $\mu_{\hat{p}} = p = 0.74$ and standard deviation $\sigma(\hat{p}) = \sqrt{\dfrac{pq}{n}} = \sqrt{\dfrac{(0.74)(0.26)}{603}} \approx 0.018$

This college has a right to brag about their retention rate. $522/603 = 86.6\%$ is over 7 standard deviations above the expected rate of 74%.

35. Polling.

Randomization condition: We must assume that the 400 voters were polled randomly.

10% condition: 400 voters polled represent less than 10% of potential voters.

Success/Failure condition: $np = 208$ and $nq = 192$ are both greater than 10.

Therefore, the sampling distribution model for $\hat{p}$ is Normal, with:

$\mu_{\hat{p}} = p = 0.52$

$\sigma(\hat{p}) = \sqrt{\dfrac{pq}{n}} = \sqrt{\dfrac{(0.52)(0.48)}{400}} \approx 0.025$

According to the Normal model, the probability that the newspaper's sample will lead them to predict defeat (that is, predict budget support below 50%) is approximately 0.212.

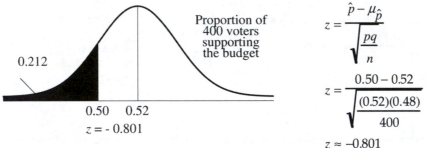

37. Apples.

Randomization condition: A random sample of 150 apples is taken from each truck.

10% condition: 150 is less than 10% of all apples.

Success/Failure Condition: $np = 12$ and $nq = 138$ are both greater than 10.

Therefore, the sampling distribution model for $\hat{p}$ is Normal, with:

$\mu_{\hat{p}} = p = 0.08$

$\sigma(\hat{p}) = \sqrt{\dfrac{pq}{n}} = \sqrt{\dfrac{(0.08)(0.92)}{150}} \approx 0.0222$

According to the Normal model, the probability that less than 5% of the apples in the sample are unsatisfactory is approximately 0.088.

Proportion of
150 apples
that are below
the standard

0.088

0.05   0.08

$z = -1.354$

$z = \dfrac{\hat{p} - \mu_{\hat{p}}}{\sqrt{\dfrac{pq}{n}}}$

$z = \dfrac{0.05 - 0.08}{\sqrt{\dfrac{(0.08)(0.92)}{150}}}$

$z \approx -1.354$

39. Nonsmokers.

Randomization condition: We will assume that the 120 customers (to fill the restaurant to capacity) are representative of all customers.

10% condition: 120 customers represent less than 10% of all potential customers.

Success/Failure condition: $np = 72$ and $nq = 48$ are both greater than 10.

Therefore, the sampling distribution model for $\hat{p}$ is Normal, with:

$$\mu_{\hat{p}} = p = 0.60$$

$$\sigma(\hat{p}) = \sqrt{\frac{pq}{n}} = \sqrt{\frac{(0.60)(0.40)}{120}} \approx 0.0447$$

Answers may vary. We will use 3 standard deviations above the expected proportion of customers who demand nonsmoking seats to be "very sure".

$$\mu_{\hat{p}} + 3\left(\sqrt{\frac{pq}{n}}\right) \approx 0.60 + 3(0.0447) \approx 0.734$$

Since $120(0.734) = 88.08$, the restaurant needs at least 89 seats in the nonsmoking section.

41. Sampling.

a) The sampling distribution model for the sample mean is $N\left(\mu, \frac{\sigma}{\sqrt{n}}\right)$.

b) If we choose a larger sample, the mean of the sampling distribution model will remain the same, but the standard deviation will be smaller.

43. Waist size.

a) The distribution of waist size of 250 men in Utah is unimodal and slightly skewed to the right. A typical waist size is approximately 36 inches, and the standard deviation in waist sizes is approximately 4 inches.

b) All of the histograms show distributions of sample means centered near 36 inches. As n gets larger the histograms approach the Normal model in shape, and the variability in the sample means decreases. The histograms are fairly Normal by the time the sample reaches size 5.

45. Waist size revisited.

a)

n	Observed mean	Theoretical mean	Observed st. dev.	Theoretical standard deviation
2	36.314	36.33	2.855	$4.019/\sqrt{2} \approx 2.842$
5	36.314	36.33	1.805	$4.019/\sqrt{5} \approx 1.797$
10	36.341	36.33	1.276	$4.019/\sqrt{10} \approx 1.271$
20	36.339	36.33	0.895	$4.019/\sqrt{20} \approx 0.897$

b) The observed values are all very close to the theoretical values.

c) For samples as small as 5, the sampling distribution of sample means is unimodal and symmetric. The Normal model would be appropriate.

d) The distribution of the original data is nearly unimodal and symmetric, so it doesn't take a very large sample size for the distribution of sample means to be approximately Normal.

47. **GPAs.**

   Randomization condition: Assume that the students are randomly assigned to seminars.

   Independence assumption: It is reasonable to think that GPAs for randomly selected students are mutually independent.

   10% condition: The 25 students in the seminar certainly represent less than 10% of the population of students.

   Large Enough Sample condition: The distribution of GPAs is roughly unimodal and symmetric, so the sample of 25 students is large enough.

   The mean GPA for the freshmen was $\mu = 3.4$, with standard deviation $\sigma = 0.35$. Since the conditions are met, the Central Limit Theorem tells us that we can model the sampling distribution of the mean GPA with a Normal model, with $\mu_{\bar{y}} = 3.4$ and standard deviation $\sigma(\bar{y}) = \dfrac{0.35}{\sqrt{25}} \approx 0.07$.

   The sampling distribution model for the sample mean GPA is approximately $N(3.4, 0.07)$.

49. **The trial of the pyx.**

   a) The mean of 100 coins varies less from sample to sample than the individual coins.

   b) $SD(\bar{y}) = \dfrac{\sigma}{\sqrt{n}} = \dfrac{0.09}{\sqrt{100}} = 0.009$

   The limit on the standard deviation of the mean should have been 0.009 grains.

Chapter 11 – Confidence Intervals for Proportions

SECTION 11.1

1.

    a) *Population* – all backpacks of concert goers; *sample* – 130 backpacks checked so far; *p* – proportion of all backpacks entering the stadium which contain alcoholic beverages; $\hat{p}$ is the proportion in the sample $\hat{p} = 17/130 = 0.1308$; yes, the confidence interval can be trusted since this seems to be a random sample.

    b) *Population* – all visitors to the English newspaper's website; *sample* – visitors to the website who responded; *p* is the proportion of all visitors to the website who approve of recent "bossnapping" actions of British workers; $\hat{p}$ is the proportion in the sample; $\hat{p} = 0.492$; no, the confidence interval should not be trusted because this is a volunteer sample and likely biased.

    c) In this case the airline is interested in knowing the weight of carry-on baggage. Weight is a quantitative (not a categorical) variable. The appropriate confidence interval would be about a mean, not a proportion. The methods of this chapter are not appropriate in this scenario.

3.

    a) $\hat{p} = \dfrac{70}{200} = 0.35$

    b) $SE(\hat{p}) = \sqrt{\dfrac{\hat{p}\hat{q}}{n}} = \sqrt{\dfrac{(0.35)(0.65)}{200}} = 0.034$

    c) $0.35 \pm (2)(0.034)$

    The 95% confidence interval is (0.282, 0.418).

SECTION 11.2

5.

    a) False. Doesn't make sense. Workers are not proportions.
    b) True.
    c) False. Our best guess is 0.48 not 0.95.
    d) False. Our best guess is 0.48, but we're not sure that's correct
    e) False. The statement should be about the true proportion, not future samples.

SECTION 11.3

7.

    a) *Population* – all customers who recently bought new cars; *sample* – 167 people surveyed about their experience; *p* –proportion of all new car buyers who are dissatisfied with the salesperson; $\hat{p}$ – proportion of new car buyers surveyed who are dissatisfied with the salesperson (3%).
    Success/Failure condition: $n\hat{p} = 167(0.03) = 5$ and $n\hat{q} = 162$. Since only 5 people were dissatisfied, the sample is **not** large enough to use a confidence interval to estimate the proportion of dissatisfied car buyers.

b) *Population* – all college students; *sample* – 2883 who were asked about their cell phones at the football stadium; p –proportion of all college students with cell phones; $\hat{p}$ – proportion of college students at the football stadium with cell phone: 2243/2883 = 77.8%.

   Plausible independence condition: whether or not a student has a cell phone shouldn't affect the probability that another does.

   Randomization condition: This sample is not random. The best we can hope for is that the students at the football stadium are representative of all college students.

   10% condition: The 2883 students at the football stadium represent less than 10% of all college students.

   The sample is large enough but extreme caution should be used when using a one-proportion z-interval to estimate the proportion of college students with cell phones. The students at the football stadium may not be representative of all students.

SECTION 11.4

9.

a) $n = \dfrac{z^2\,\hat{p}\hat{q}}{ME^2} = \dfrac{(1.96)^2(0.5)(0.5)}{(0.02)^2} = 2401$

   2401 students should be sampled in order to be 95% confident that the estimate is within 0.02 of the true proportion.

b) $n = \dfrac{z^2\,\hat{p}\hat{q}}{ME^2} = \dfrac{(2.576)^2(0.5)(0.5)}{(0.02)^2} = 4147.36$

   4148 students should be sampled in order to be 99% confident that the estimate is within 0.02 of the true proportion.

c) $n = \dfrac{z^2\,\hat{p}\hat{q}}{ME^2} = \dfrac{(1.96)^2(0.5)(0.5)}{(0.05)^2} = 384.16$

   385 students should be sampled in order to be 95% confident that the estimate is within 0.05 of the true proportion.

11.

a)

$$ME = z^* \sqrt{\dfrac{\hat{p}\hat{q}}{n}}$$

$$0.06 = 1.645\sqrt{\dfrac{(0.25)(0.75)}{n}}$$

$$n = \dfrac{(1.645)^2(0.25)(0.75)}{(0.06)^2}$$

$$n \approx 141 \text{ people}$$

In order to estimate the proportion of non-graduates in the 25-to 30-year-old age group to within 6% with 90% confidence, we would need a sample of at least 141 people. All decimals in the final answer must be rounded up, to the next person.

(For a more cautious answer, let $\hat{p} = \hat{q} = 0.5$. This method results in a required sample of 188 people.)

b)

$$ME = z^* \sqrt{\frac{\hat{p}\hat{q}}{n}}$$

$$0.04 = 1.645 \sqrt{\frac{(0.25)(0.75)}{n}}$$

$$n = \frac{(1.645)^2 (0.25)(0.75)}{(0.04)^2}$$

$$n \approx 318 \text{ people}$$

In order to estimate the proportion of non-graduates in the 25-to 30-year-old age group to within 4% with 90% confidence, we would need a sample of at least 318 people. All decimals in the final answer must be rounded up, to the next person.

(For a more cautious answer, let $\hat{p} = \hat{q} = 0.5$. This method results in a required sample of 423 people.)

Alternatively, the margin of error is now 2/3 of the original, so the sample size must be increased by a factor of 9/4. $141(9/4) \approx 318$ people.

c)

$$ME = z^* \sqrt{\frac{\hat{p}\hat{q}}{n}}$$

$$0.03 = 1.645 \sqrt{\frac{(0.25)(0.75)}{n}}$$

$$n = \frac{(1.645)^2 (0.25)(0.75)}{(0.03)^2}$$

$$n \approx 564 \text{ people}$$

To estimate the proportion of non-graduates in the 25-to 30-year-old age group to within 3% with 90% confidence, we would need a sample of at least 564 people. All decimals in the final answer must be rounded up, to the next person.

(For a more cautious answer, let $\hat{p} = \hat{q} = 0.5$. This method results in a required sample of 752 people.)

Alternatively, the margin of error is now half that of the original, so the sample size must be increased by a factor of 4. $141(4) \approx 564$ people.

CHAPTER EXERCISES

13. Margin of error.

He believes the true percentage of voters with a certain opinion is within 4% of his estimate with some degree of confidence, probably 95%. The margin of error is often stated without a confidence level and, most of the time, it can be assumed that the level is at 95%.

15. Conditions.

a) *Population* –all cars in the local area; *sample* –134 cars actually stopped at the checkpoint; *p* –population proportion of all cars with safety problems; $\hat{p}$ - proportion of cars in the sample that actually have safety problems (10.4%).

Plausible independence condition: there is no reason to believe that the safety problems of cars are related to each other.

Randomization condition: this sample is not random, so hopefully the cars stopped are representative of cars in the area.

10% condition: the 134 cars stopped represent a small fraction of all cars in the local area, certainly less than 10%.

Success/Failure condition: $n\hat{p} = 14$ and $n\hat{q} = 120$ are both greater than 10, so the sample is large enough.

A one proportion *z*-interval can be calculated for the proportion of all cars in the area with safety problems.

b) *Population* –the general public; *sample* –602 viewers that logged on to the Web site; *p* –population proportion of the general public that think corporate corruption is "worse";

$\hat{p}$ –proportion of viewers that logged on to the Web site and voted that corporate corruption is "worse" (81.1%).

Randomization condition: This sample is not random, but biased by voluntary response. It would be very unwise to attempt to use this sample to infer anything about the opinion of the general public related to corporate corruption.

17. Catalog sales. If repeated samples of the same sample size were taken and confidence intervals constructed (according to the formula given in the text), in the long run 95% of these intervals would contain the true on-time arrival rate.

    a) Not correct. This is not the meaning of a confidence interval.

    b) Not correct. Different samples will give different results. Most likely, none of the samples will have *exactly* 88% on-time orders.

    c) Not correct. A confidence interval says something about the unknown population proportion, not the sample proportion in different samples.

    d) Not correct. In this sample, we *know* that 88% arrived on time. There is no need to make an interval for the sample proportion.

    e) Not correct. The interval is about the proportion of on-time orders, not about the days.

19. Confidence intervals.

    a) False. For a given sample size, higher confidence means a *larger* margin of error because we are less sure of falling in a smaller interval.

    b) True. Larger samples lead to smaller standard errors, which lead to smaller margins of error.

    c) True. Larger samples are less variable, which makes us more confident that a given confidence interval succeeds in catching the population proportion.

    d) False. The margin of error decreases as the square root of the sample size increases. Halving the margin of error requires a sample four times as large as the original.

21. Cars. We are 90% confident that between 29.9% and 47.0% of U.S. cars are made in Japan.

23. E-mail.

    a) $ME = z^* \times SE(\hat{p}) = z^* \sqrt{\dfrac{\hat{p}\hat{q}}{n}} = 1.645 \times \sqrt{\dfrac{(0.38)(0.62)}{1012}} \approx 0.025 = 2.5\%$

    b) The pollsters are 90% confident that the true proportion of adults who do not use e-mail is within 2.5% of the estimated 38%.

    c) A 99% confidence interval requires a larger margin of error. In order to increase confidence, the interval must be wider

    d) $ME = z^* \times SE(\hat{p}) = z^* \sqrt{\dfrac{\hat{p}\hat{q}}{n}} = 2.576 \times \sqrt{\dfrac{(0.38)(0.62)}{1012}} \approx 0.039 = 3.9\%$

    e) Smaller margins of error will give us less confidence in the interval.

25. Teenage drivers.

    a) **Plausible independence condition:** There is no reason to believe that accidents selected at random would be related to one another.
       Randomization condition: The insurance company randomly selected 582 accidents.
       10% condition: 582 accidents represent less than 10% of all accidents.

Success/Failure condition: $n\hat{p} = 91$ and $n\hat{q} = 491$ are both greater than 10, so the sample is large enough.

Since the conditions are met, we can use a one-proportion z-interval to estimate the percentage of accidents involving teenagers.

$$\hat{p} \pm z^* \sqrt{\frac{\hat{p}\hat{q}}{n}} = \left(\frac{91}{582}\right) \pm 1.960\sqrt{\frac{\left(\frac{91}{582}\right)\left(\frac{491}{582}\right)}{582}} = (12.7\%, 18.6\%)$$

b) We are 95% confident that between 12.7% and 18.6% of all accidents involve teenagers.

c) About 95% of random samples of size 582 will produce intervals that contain the true proportion of accidents involving teenagers.

d) Our confidence interval contradicts the assertion of the politician. The figure quoted by the politician, 1 out of every 5, or 20%, is outside the interval.

27. Retailers. The grocer can conclude nothing about the opinions of all his customers from this survey. Those customers who bothered to fill out the survey represent a voluntary response sample, consisting of people who felt strongly one way or another about irradiated food. The random condition was not met.

29. Internet music. This was a random sample of less than 10% of all Internet users; there were 703 x 0.18 = 127 successes and 576 failures, both at least 10.

The 95% confidence interval: $\hat{p} \pm z^* \sqrt{\frac{\hat{p}\hat{q}}{n}} = 0.18 \pm 1.96\sqrt{\frac{(0.18)(0.82)}{703}} = 0.18 \pm 0.0284 = (0.152, 0.208)$. We

are 95% confident that between 15.2% and 20.8% of Internet users have downloaded music from a site that was not authorized.

31. International business.

   a) 385/550 = 0.70; 70% of U.S. chemical companies in the sample are certified.

   b) This was a random sample, but we don't know if it is less than 10% of all U.S. chemical companies; there were 550 x 0.70 = 385 successes and 165 failures, both at least 10. The 95%

    confidence interval: $\hat{p} \pm z^* \sqrt{\frac{\hat{p}\hat{q}}{n}} = 0.70 \pm 1.96\sqrt{\frac{(0.70)(0.30)}{550}} = 0.70 \pm 0.038 = (0.662, 0.738)$. We

    are 95% confident that between 66.2% and 73.8% of the chemical companies in the U.S. are certified. It appears that the proportion of companies certified in the U.S. is less than in Canada.

33. Business ethics.

   a) There may be response bias based on the wording of the question.

   b) The approval proportions: $0.53*538 + 0.44*538 = 521.86$ or $522/(538+538) = 0.485$. The 95% confidence interval:

$$\hat{p} \pm z^* \sqrt{\frac{\hat{p}\hat{q}}{n}} = 0.485 \pm 1.96\sqrt{\frac{(0.485)(0.515)}{(1076)}} = 0.485 \pm 0.030 = (0.455, 0.515)$$

   c) The margin of error based on the pooled sample is smaller, since the sample size is larger.

35. Gambling.

   a) The interval based on the survey conducted by the college Statistics class will have the larger margin of error, since the sample size is smaller.

   b) **Plausible independence condition:** There is no reason to believe that one randomly selected voter's response will influence another.

Randomization condition: Both samples were random.
10% condition: Both samples are probably less than 10% of the city's voters, provided the city has more than 12,000 voters.
Success/Failure condition:
For the newspaper, $n_1\hat{p}_1 = (1200)(0.53) = 636$ and $n_1\hat{q}_1 = (1200)(0.47) = 564$
For the Statistics class, $n_2\hat{p}_2 = (450)(0.54) = 243$ and $n_2\hat{q}_2 = (450)(0.46) = 207$
All the expected successes and failures are greater than 10, so the samples are large enough.

Since the conditions are met, we can use one-proportion z-intervals to estimate the proportion of the city's voters that support the gambling initiative.

Newspaper poll: $\hat{p}_1 \pm z^* \sqrt{\dfrac{\hat{p}_1\hat{q}_1}{n_1}} = (0.53) \pm 1.960 \sqrt{\dfrac{(0.53)(0.47)}{1200}} = (50.2\%, 55.8\%)$

Statistics class poll: $\hat{p}_2 \pm z^* \sqrt{\dfrac{\hat{p}_2\hat{q}_2}{n_2}} = (0.54) \pm 1.960 \sqrt{\dfrac{(0.54)(0.46)}{450}} = (49.4\%, 58.6\%)$

a) The Statistics class should conclude that the outcome is too close to call, because 50% is in their

37. Pharmaceutical company.

a) **Plausible independence condition:** It is reasonable to think that the randomly selected children are mutually independent in regards to vitamin D deficiency.
Randomization condition: The 2,700 children were chosen at random.
10% condition: 2,700 children are less than 10% of all English children.
Success/Failure condition: $n\hat{p} = (2,700)(0.20) = 540$ and $n\hat{q} = (2,700)(0.80) = 2160$ are both greater than 10, so the sample is large enough.

Since the conditions are met, we can use a one-proportion z-interval to estimate the proportion of the English children with vitamin D deficiency.

$\hat{p} \pm z^* \sqrt{\dfrac{\hat{p}\hat{q}}{n}} = (0.20) \pm 2.326 \sqrt{\dfrac{(0.20)(0.80)}{2700}} = (18.2\%, 21.8\%)$

b) We are 98% confident that between 18.2% and 21.8% of English children are deficient in vitamin D.

c) About 98% of random samples of size 2700 will produce confidence intervals that contain the true proportion of English children that are deficient in vitamin D.

d) No, the interval says nothing about causation.

39. Funding.

a) This is not a random sample; even though it is representative, we cannot be sure if we have less than 10% of all WBCs; there are 8 successes which are not greater than 10, so the sample is not large enough.

b) Since the conditions are not met but the sample is representative, we could perform a pseudo observation confidence interval, adding two successes and two failures to the data. The resulting 90% confidence interval based on 10 successes out of 24:

$\hat{p} \pm z^* \sqrt{\dfrac{\hat{p}\hat{q}}{n}} = 0.4167 \pm 1.645 \sqrt{\dfrac{(0.417)(0.583)}{(24)}} = 0.4167 \pm 0.1656 = (0.251, 0.582) = (25.1\%, 58.2\%)$

41. Benefits survey.
  a) This was a random sample of less than 10% of all companies in Vermont; there were 12 successes and zero failures, which is not greater than 10, so the sample is not large enough.
  b) The pseudo 90% confidence interval based on 14 successes out of 16:

$$\hat{p} \pm z * \sqrt{\frac{\hat{p}\hat{q}}{n}} = 0.875 \pm 1.96 \sqrt{\frac{(0.875)(0.125)}{(16)}} = 0.875 \pm 0.162 = (0.713, 1.037)$$

  We can say that with 95% confidence the true proportion is at least 71.3%.

43. IRS.
  a) This was a random sample of less than 10% of all self-employed taxpayers; there were 20 successes and 206 failures, both at least 10.

  b) The 95% confidence interval:

$$\hat{p} \pm z * \sqrt{\frac{\hat{p}\hat{q}}{n}} = 0.088 \pm 1.96 \sqrt{\frac{(0.0885)(0.9115)}{226}} = 0.088 \pm 0.037 = (0.051, 0.125) = (5.1\%, 12.5\%)$$

  c) We are 95% confident that between 5.1% and 12.6% of all self-employed individuals had their tax returns audited in the past year.

  d) If we were to select repeated samples of 226 individuals, we'd expect about 95% of the confidence intervals we created to contain the true proportion of all self-employed individuals who were audited.

45. Internet music, again.
  a) This was a random sample of less than 10% of all U.S. adults; there were 703 * 0.13 = 91 successes and 612 failures, both at least 10.

  b) The 95% confidence interval:

$$\hat{p} \pm z * \sqrt{\frac{\hat{p}\hat{q}}{n}} = 0.13 \pm 1.96 \sqrt{\frac{(0.13)(0.87)}{(703)}} = 0.13 \pm 0.025 = (0.105, 0.155) = (10.5\%, 15.5\%)$$

47. Politics.
  a) The confidence interval for the true proportion of all 18- to 29-year olds who believe the U.S. is ready for a woman president will be about twice as wide as the confidence interval for the true proportion of all U.S. adults, since it is based on a sample about one-fourth as large (assuming approximately equal proportions).

  b) **Plausible independence condition:** The responses are likely to be independent.
      Randomization condition: The respondents were randomly selected.
      10% condition: 250 adults are less than 10% of all U.S. adults.
      Success/Failure condition: $n\hat{p} = 250(0.62) = 155$ and $n\hat{q} = 250(0.38) = 95$ are both greater than 10, so the sample is large enough.

  The 95% confidence interval:

$$\hat{p} \pm z * \sqrt{\frac{\hat{p}\hat{q}}{n}} = 0.62 \pm 1.96 \sqrt{\frac{(0.62)(0.38)}{(250)}} = 0.62 \pm 0.060 = (0.560, 0.680) = (56.0\%, 68.0\%)$$

  We are 95% confident that between 56.0% and 68.0% of 18- to 29-year olds believe that the U.S. is ready for a woman president.

49. More internet music.

   a) The parameter is the proportion of digital songs in student libraries that are legal. The population is all songs held in digital libraries. The sample size is 117,079 songs, not 168 students.

   b) **Plausible independence condition:** The songs and whether or not they are paid for are likely to be independent.

      Randomization condition: The sample is random, but this is a cluster sample. We have information about 168 clusters of songs in the 168 digital music libraries we asked about.

      10% condition: 117, 079 is much less than 10% of all digital songs.

      Success/Failure condition: The number of legal songs and illegal songs in the sample are both much greater than 10, so the sample is large enough.

      Since the conditions are satisfied, we can use a one-proportion z-interval to estimate the proportion of digital songs that are legal

      The 95% confidence interval:

$$\hat{p} \pm z^* \sqrt{\frac{\hat{p}\hat{q}}{n}} = 0.231 \pm 1.96\sqrt{\frac{(0.231)(0.769)}{(117,079)}} = 0.231 \pm 0.0024 = (0.2286, 0.2334) = (22.9\%, 23.3\%)$$

   c) We are 95% confident that between 22.9% and 23.3% of digital songs were legally purchased.

   d) We can be confident – the very large sample size means that we should be very certain about the results – thus the narrow confidence interval.

51. CDs.

   a) This was a random sample of less than 10% of all Internet users; there were 703* 0.64 = 450 successes and 253 failures, both at least 10. A 90% confidence interval:

$$\hat{p} \pm z^* \sqrt{\frac{\hat{p}\hat{q}}{n}} = 0.64 \pm 1.645\sqrt{\frac{(0.64)(0.36)}{(703)}} = 0.64 \pm 0.030 = (0.610, 0.670) = (61.0\%, 67.0\%)$$

      We are 90% confident that between 61.0% and 67.0% of Internet users would still buy a CD.

   b) In order to cut the margin of error in half, they must sample 4 times as many users because the sample size is inside the square root and in the denominator.

$$ME = z^* \sqrt{\frac{\hat{p}\hat{q}}{n}}$$ The square root of ¼ = ½, therefore, n must be multiplied by 4;

      4 x 703 = 2812 users.

53. Graduation.

$$ME = z^* \sqrt{\frac{\hat{p}\hat{q}}{n}}$$

$$0.02 = 1.960\sqrt{\frac{(0.25)(0.75)}{n}}$$

$$n = \frac{(1.960)^2 (0.25)(0.75)}{(0.02)^2}$$

$$n \approx 1{,}801 \text{ people}$$

In order to estimate the proportion of non-graduates in the 25-to 30-year-old age group to within 2% with 95% confidence, we would need a sample of at least 1,801 people. All decimals in the final answer must be rounded up, to the next person.

(For a more cautious answer, let $\hat{p} = \hat{q} = 0.5$. This method results in a required sample of 2,401 people.)

55. Pilot study.

$$ME = z^* \sqrt{\frac{\hat{p}\hat{q}}{n}}$$

$$0.03 = 1.645 \sqrt{\frac{(0.15)(0.85)}{n}}$$

$$n = \frac{(1.645)^2 (0.15)(0.85)}{(0.03)^2}$$

$$n \approx 384 \text{ cars}$$

Use $\hat{p} = \frac{9}{60} = 0.15$ from the pilot study as an estimate.

In order to estimate the percentage of cars with faulty emissions systems to within 3% with 90% confidence, the state's environmental agency will need a sample of at least 384 cars. All decimals in the final answer must be rounded up, to the next car.

57. Approval rating.

$$ME = z^* \sqrt{\frac{\hat{p}\hat{q}}{n}}$$

$$0.025 = z^* \sqrt{\frac{(0.65)(0.35)}{972}}$$

$$z^* = \frac{0.025}{\sqrt{\frac{(0.65)(0.35)}{972}}}$$

$$z^* \approx 1.634$$

Since $z^* \approx 1.634$, which is close to 1.645, the pollsters were probably using 90% confidence. The slight difference in the z^* values is due to rounding of the governor's approval rating.

59. Customer spending.

This was a random sample of less than 10% of all customers. There were 67 successes and 433 failures, both at least 10. From the data set, $\hat{p} = 67/500 = 0.134$.

The 95% confidence interval:

$$\hat{p} \pm z^* \sqrt{\frac{\hat{p}\hat{q}}{n}} = 0.134 \pm 1.96 \sqrt{\frac{(0.134)(0.866)}{(500)}} = 0.134 \pm 0.030 = (0.104, 0.164) = (10.4\%, 16.4\%)$$

We are 95% confident that the true proportion of customers that spend $1000 per month or more is between 10.4% and 16.4%.

61. Health insurance.

a. This was a random sample of less than 10% of all MA males. There were 2662 successes and 398 failures, both at least 10.

b. The 95% confidence interval:

$$\hat{p} \pm z^* \sqrt{\frac{\hat{p}\hat{q}}{n}} = 0.87 \pm 1.96 \sqrt{\frac{(0.87)(0.13)}{(3060)}} = 0.87 \pm 0.012 = (0.858, 0.882) = (85.8\%, 88.2\%)$$

c. We are 95% confident that between 85.8% and 88.2% of MA males has health insurance.

Chapter 12 – Confidence Intervals for Means

SECTION 12.1

1.

a) $SE(\bar{y}) = \dfrac{s}{\sqrt{n}} = \dfrac{9.84}{\sqrt{25}} = 1.968$ years

b) $SE(\bar{y}) = \dfrac{s}{\sqrt{n}} = \dfrac{9.84}{\sqrt{100}} = 0.984$ years

Changing the sample size from 25 to 100 has cut the standard error in half.

3.

a) The degrees of freedom for the t-statistic is $n-1 = 25 - 1 = 24$

b) The degrees of freedom for the t-statistic is $n-1 = 100 - 1 = 99$

SECTION 12.2

5.

a) In Table T use $df = 24$ and Two-tail probability of 0.05 to get $t^* = 2.064$.

b) In Table T use $df = 99$ and Two-tail probability of 0.05 to get $t^* = 1.984$.

7.

a) To find the confidence interval use $\bar{y} \pm t^*_{n-1} \times SE(\bar{y})$ where $SE(\bar{y}) = \dfrac{s}{\sqrt{n}}$:

$31.84 \pm (2.064)\dfrac{9.84}{\sqrt{25}}$

The 95% confidence interval for the mean age is (27.78, 35.90) years.

b) The margin of error is $(2.064)\dfrac{9.84}{\sqrt{25}} = 4.06$ years.

c) If the standard deviation was known to be 10, then we have

$31.84 \pm (1.96)\dfrac{10}{\sqrt{25}}$

and the 95% confidence interval would be (27.92, 35.76) years which is slightly narrower.

SECTION 12.3

9. Independence: The data were from a random survey and should be independent

Randomization: the data were selected randomly.

10% Condition: These customers are fewer than 10% of the customer population.

Nearly Normal: The histogram is unimodal and symmetric, which is sufficient.

SECTION 12.5

11.

a) Cut the margin of error which is 4.06 years in half to 2.03 years and find n.

$\sqrt{n} = \dfrac{(2.064)(9.84)}{2.03}$ so $n = (10.005)^2 \approx 100$

b) Cut the margin of error which is 4.06 years by 10 to 0.406 years and find n.

$$\sqrt{n} = \frac{(2.064)(9.84)}{0.406} \text{ so } n = (50.024)^2 \approx 2500$$

CHAPTER EXERCISES

13. *t*-**models.** The critical t values for confidence intervals are defined as the cutoff values for specific degrees of freedom (n–1) for a specific confidence level. Using technology or tables:

    a) 1.74

    b) 2.37

15. Confidence intervals. As the variability of a sample increases, the width of a 95% confidence interval increases, assuming the sample size remains the same. The standard error is proportional to the variability

    (s) of the sample ($\frac{s}{\sqrt{n}}$).

17. Confidence intervals and sample size.

    a) The confidence interval equation is: $\bar{y} \pm t^{*}_{n-1} \times SE(\bar{y})$ where $SE = \frac{s}{\sqrt{n}}$. We need to find the critical t value before solving the equation. This is found through technology or a table for a confidence level of 95% at degrees of freedom (*df*) of 29 (n-1). Substituting in the equation:

$$\$4.49 \pm 2.045 * \frac{0.29}{\sqrt{30}} = \$4.49 \pm 0.1083 = (\$4.382, \$4.598) .$$

    b) The confidence interval equation is: $\bar{y} \pm t^{*}_{n-1} \times SE(\bar{y})$ where $SE = \frac{s}{\sqrt{n}}$. We need to find the critical t value before solving the equation. This is found through technology or a table for a confidence level of 95% at degrees of freedom (*df*) of 29 (n-1). Substituting in the equation:

$$\$4.49 \pm 1.7 * \frac{0.29}{\sqrt{30}} = \$4.49 \pm 0.0900 = (\$4.400, \$4.580) .$$

    c) The confidence interval at a 95% confidence level for $n = 60$ is:

$$\$4.49 \pm 2.0 * \frac{0.29}{\sqrt{60}} = \$4.49 \pm 0.075 = (\$4.415, \$4.565)$$

19. Marketing livestock feed.

    a) This interpretation is not correct. A confidence interval addresses the mean weight gain of the population of all cows. It says nothing about individual cows.

    b) This interpretation is not correct. A confidence interval addresses the mean weight gain of the population of all cows. It says nothing about individual cows.

    c) This interpretation is not correct. We are certain the mean weight gain of the cows in the study is 56 pounds.

    d) This interpretation is not correct. The statement implies that the average weight gain varies, which it doesn't.

    e) This interpretation is not correct. The statement implies that there is something special about our interval. This interval is actually one of many that could have been generated, depending on the cows chosen for the sample.

21. CEO compensation. The assumptions and conditions for a *t*-interval are not satisfied. For a sample size of only 20, the distribution is clearly skewed to the right. There is a large outlier pulling the mean to a higher value.

23. Parking.
   a)   The assumptions required in order to use these statistics for inference are that the data are a random sample of all days. In addition, the distribution should be unimodal and symmetric with no outliers.

   b)   The confidence interval equation is: $\bar{y} \pm t^{*}_{n-1} \times SE(\bar{y})$ where $SE = \dfrac{s}{\sqrt{n}}$. Substituting values:

   $$\$126 \pm 1.68 * \frac{15}{\sqrt{44}} = \$126 \pm 3.80 = (\$122.20, \$129.80).$$

   c)   We are 90% confident that the interval ($122.20, $129.80) contains the true mean daily income of the parking garage.

   d)   90% of all random sample of size 44 will produce intervals that contain the true mean daily income of the parking garage.

   e) $128 is within the interval. It is a plausible value.

25. Parking, part 2.
   a)   The 95% confidence level gives us increased confidence that the mean parking revenue is contained with the interval.

   a)   The increased confidence level widens the confidence interval due to higher error. The resulting interval is wider than before.

   b)   By collecting a larger sample, they can decrease the error and, therefore, decrease the margin of error and the resulting confidence interval.

27. State budgets.

   a)   The confidence interval equation is: $\bar{y} \pm t^{*}_{n-1} \times SE(\bar{y})$ where $SE = \dfrac{s}{\sqrt{n}}$.

   Substituting values: $\$2350 \pm 2.009 * \dfrac{425}{\sqrt{51}} = \$2350 \pm 119.56 = (\$2230.44, \$2469.56).$

   b)   The assumptions required are independence – probably OK, randomization not specified and sample probably not more than 10% of population. The next assumption is the normal population assumption – check by looking at a histogram of the data. With a sample size of 49, the *t* procedures should be acceptable to use as long as the distribution is at least somewhat symmetric and unimodal.
   c)   We are 95% confident that the interval $2227.93 to $2472.07 contains the true mean increase in sales tax revenue.

29. Departures.
   a)   The assumptions required are independence – OK, the monthly on-time departure rates should be independent. Although this is not a random sample, the chosen months should be representative, and the sample is fewer than 10% of all months. The next assumption is the normal population assumption – check by looking at a histogram of the data. The histogram looks unimodal but is slightly left-skewed. This is not of concern due to the large sample size.

   b)   The confidence interval equation is: $\bar{y} \pm t^{*}_{n-1} \times SE(\bar{y})$ where $SE = \dfrac{s}{\sqrt{n}}$. Substituting values:

   $$81.1838 \pm 1.65558 * \frac{4.47094}{\sqrt{144}} = 81.1838 \pm 0.6168 = (80.57, 81.80) \text{ or } 80.57\% \text{ to } 81.80\% \text{ OT departures.}$$

c)   We can be 90% confident that the interval from 80.57% to 81.80% holds the true mean monthly percentage of on-time flight departures.

31. Computer lab fees.
   a)   The histogram of lab times shows 2 extreme outliers that are outside of the majority of data points. There are large gaps between the majority of the times and the outliers, one low and one high. Because of the outliers, the conditions for inference are violated.

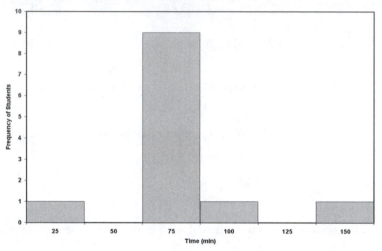

   b)

   To find the confidence interval use: $\bar{y} \pm t^{*}_{n-1} \times SE(\bar{y})$ where $SE(\bar{y}) = \dfrac{s}{\sqrt{n}}$

   $$63.25 \pm (2.201)\dfrac{28.926}{\sqrt{12}} = 63.25 \pm 18.379$$

   The 95% confidence for the true mean (44.871, 81.629) includes the previous mean time of 55 minutes. Since 55 is a plausible value for the mean, there is no evidence to support the claim. We would be reluctant to conclude that the mean time has increased to above 55 minutes.

   The confidence interval without the outliers is $61.5 \pm (2.262)\dfrac{9.595}{\sqrt{10}} = 61.5 \pm 6.863$

   The 95% confidence interval for the true mean (54.637, 68.363) without the outliers also includes the previous mean time of 55 minutes (just barely).

33. Growth and air pollution.
   a)   The assumptions and conditions that must be satisfied are: the data comes from a nearly normal distribution (independent air samples). The air samples were selected randomly and there is no bias present in the sample.

   b)   The histogram of air samples is not exactly normal but the sample size is large so inference is OK.

35. Traffic speed.
   a)   The upper limit on the confidence interval is 14.9 and the mean is 11.6. Thus the margin of error is 14.9 – 11.6 = 3.3 mph.

b) To reduce the margin of error, the sample size can be increased. The required sample size can be calculated from the equation: $ME = t^*_{n-1} \times \dfrac{s}{\sqrt{n}} = 1.96 \times \dfrac{8}{\sqrt{n}} = 2$. Solve for n:

$$\sqrt{n} = 7.84 \; ; \; n = (7.84)^2 = 61.466 = 62.$$

37. Tax audits.

a) The confidence interval equation is: $\bar{y} \pm t^*_{n-1} \times SE(\bar{y})$ where $SE = \dfrac{s}{\sqrt{n}}$. Substituting values:

$$\$680 \pm 2.04 * \dfrac{75}{\sqrt{32}} = \$680 \pm 27.05 = (\$653, \$707).$$

b) The confidence interval does not contain the hypothesized mean of $650. This provides evidence that the current year's mean audit cost is higher than $650.

39. Wind power.

a) The timeplot shows no pattern which means that the measurements are independent from each other. Although this is not a random sample, an entire year is measure, so it is likely that we have representative values. The sample is fewer than 10% of all possible wind readings. Both the histogram and the normal probability plot indicate that the distribution is near normality. We can proceed with the inferences procedures.

b) Find the 95% confidence interval for the annual average wind speed using

$$\bar{y} \pm t^*_{n-1} \times SE(\bar{y}) = 8.019 \pm 1.96 \times 0.114$$

$$= 8.019 \pm 0.224$$

The 95% confidence interval is 7.795 to 8.243 mph. Since the required annual average wind speed should be at least 8 mph in order to generate enough energy, and this interval includes values below 8 mph, you would not recommend building the turbine at this site.

Chapter 13 – Testing Hypotheses

SECTION 13.1

1.

a) Let p = proportion of wins on the slot machine.
H_0: $p = 0.01$ vs. H_A: $p \neq 0.01$.

b) Let μ = mean amount spent per visit by customers on the company's website.
H_0: $\mu = \$35.32$ vs. H_A: $\mu \neq \$35.32$.

c) Let p = proportion of patients cured by the new drug.
H_0: $p = 0.30$ vs. H_A: $p \neq 0.30$.

d) Let p = proportion of clients now using the bank's website.
H_0: $p = 0.40$ vs. H_A: $p \neq 0.40$.

SECTION 13.3

3.

a) False. A very high P-value shows that the data are consistent with the null hypothesis; it does not provide evidence for rejecting the null hypothesis.

b) False. A very low P-value provides evidence for rejecting the null hypothesis, but it does not *prove* that it is false.

c) False. A high P-value shows that the data are consistent with the null hypothesis, but it does not *prove* that the null hypothesis is true.

d) False. Whether a P-value provides enough evidence to reject the null hypothesis depends on the risk of a type I error that one is willing to assume (the alpha level). This value is set to be quite low. A typical alpha level commonly used is 0.05.

SECTION 13.4

5.

a) $SE(\hat{p}) = \sqrt{\dfrac{p_0 q_0}{n}} = \sqrt{\dfrac{(0.35)(0.65)}{300}} = 0.028$

b) $z = \dfrac{\hat{p} - p_0}{SD(\hat{p})} = \dfrac{0.46 - 0.35}{0.028} = 3.93$

c) Yes, the z-statistic is unusually large as it very unlikely to see values above 3.

SECTION 13.5

7.

a) H_0: $p = 0.40$ vs. H_A: $p \neq 0.40$. This is a two-sided alternative.

b) H_0: $\mu = 5.4$ minutes vs. H_A: $\mu < 5.4$ minutes. This is one-sided alternative.

c) H_0: $p = 0.42$ vs. H_A: $p > 0.42$. This is a one-sided alternative.

d) H_0: $p = 0.50$ vs. H_A: $p > 0.50$. This is a one-sided alternative.

SECTION 13.6

9.

a) H_0: $\mu = 25$ years.

b) H_A: $\mu \neq 25$ years. This is a two-sided alternative.

c)   To find the value of the test statistic use $t_{n-1} = \dfrac{\bar{y} - \mu_0}{SE(\bar{y})}$.

$$t_{24} = \frac{31.84 - 25}{1.968} = 3.476$$

d)   Using Table T, we see that the P-value < 0.001 (with technology we find that it is 0.000977).

e)   Since the P-value < α = 0.05, we reject H_0. There is strong evidence that the mean age is not 25.

f)   Because the 95% confidence interval does not contain 25, we can reject H_0 at α = 0.05.

SECTION 13.7

11.

a)   True.

b)   False. The alpha level is set independently and does not depend on the sample size.

c)   False. The P-value would have to be less than 0.01 to reject the null hypothesis.

d)   False. It simply means we do not have enough evidence at that alpha level to reject the null hypothesis.

SECTION 13.8

13.

a)   Since this is a test for one proportion, we use z. Since there is a two sided alternative, we place $\alpha/2$ or 0.025 in each tail of the standard normal curve. Looking up the probabilities of 0.025 and 0.975 in Table Z, we find $z = \pm1.96$.

b)   Since this is a test for one proportion, we use z. Since this is a one-sided alternative, we place α = 0.05 in the right tail of the standard normal curve. Looking up the probability 0.95 in Table Z we find z = 1.645.

c)   Since this is a test for one mean, use t with $n-1$ or 35 df. Since this is a two-sided alternative, we place $\alpha/2$ or 0.025 in each tail of the t-distribution. Looking up the two-tail probability of 0.05 and 35 df we find $t = \pm2.030$

d)   Since this is a test for one proportion, we use z. Since this is a one-sided alternative, we place α = 0.01 in the right tail of the standard normal curve. Looking up the probability 0.99 in Table Z we find z = 2.33. Note that n is not relevant for critical values of z.

e)   Since this is a test for one mean, use t with $n-1$ or 999 df. Note that with such a large sample, the values for critical values for t are identical to critical values for z. Since this is a one-sided alternative, we place α = 0.01 in the left tail of the distribution. Looking up the probability of 0.01 in Table Z we find z = -2.33.

SECTION 13.9

15.

a)   Use $\hat{p} \pm 1.96 SE(\hat{p})$

0.25 ± 1.96×0.0274
(0.196, 0.304)

b)   No, because 0.20 is a plausible value as it is contained in the interval.

c)   The SE is based on $\hat{p}$ $SE(\hat{p}) = \sqrt{\dfrac{\hat{p}\hat{q}}{n}} = \sqrt{\dfrac{(0.25)(0.75)}{250}} = 0.0274$. The SD is based on the hypothesized value 0.20, so $SE(\hat{p}) = \sqrt{\dfrac{p_0 q_0}{n}} = \sqrt{\dfrac{(0.20)(0.80)}{250}} = 0.0253$

d) The SE is used in constructing the confidence interval since it is sample based.

17.

a) $SE = \dfrac{s}{\sqrt{n}} = \dfrac{2.0}{\sqrt{25}} = 0.4$

b) t^* with 24 df for a 90% confidence interval is 1.711

c) Use $\bar{y} \pm t^* SE(\bar{y})$

$16.5 \pm 1.711 \times 0.4$
$(15.82, 17.18)$

d) No, we fail to reject H_0 (one-sided) at $\alpha = 0.05$ because 16.0 is contained in the corresponding two-sided 90% confidence interval.

SECTION 13.10

19.

a) Type I error. The true value of customers enrolled on the bank's website is not greater than 0.3, yet the bank rejected the null hypothesis that was true.

b) No error has been committed. The true proportion of students preferring Coke is 0.50, and the null hypothesis was not rejected. A correct conclusion was made.

c) Type II error. The true score of this year's applicants on the placement exam was 55.3 points which is greater than the average score attained by last year's applicants (52.5). She failed to reject the null hypothesis that was false.

d) Type II error. The true relief rate was 0.38 which was greater than 0.25. The pharmaceutical company did not reject the null hypothesis that was false.

CHAPTER EXERCISES

21. Hypotheses.
    a) Let p be the proportion of products delivered on time (0.90). This is stated in the null hypothesis. We are testing to see if the proportion of products delivered on time meets the standard of at least 90%. This defines the alternative hypothesis. Percentages are converted to proportions.
        $H_0: p = 0.90$
        $H_A: p > 0.90$

    b) Let p be the proportion of houses taking more than 3 months to sell (0.50). This is stated in the null hypothesis. The problem states that a realty company says that the percentage is now > 50%. This defines the alternative hypothesis. Percentages are converted to proportions.
        $H_0: p = 0.50$
        $H_A: p > 0.50$

    c) Let p be the error rate proportion (0.02). This is stated in the null hypothesis. The problem states that the reports have a percentage that is < 2%. This defines the alternative hypothesis. Percentages are converted to proportions.
        $H_0: p = 0.02$
        $H_A: p < 0.02$

23. Deliveries. Statement d) is the correct interpretation of the P-value.

25. P-value. If the rate of seat belt usage after the campaign is the same as the rate of seat belt usage before the campaign, there is a 17% chance of observing a rate of seat belt usage this large or larger after the campaign in a sample of the same size by natural sampling variation alone.

27. Ad campaign. Statement e) is the correct interpretation of the P-value.

29. Product effectiveness. It is not reasonable to conclude that the new formula and the old one are equally effective. Furthermore, our inability to make that conclusion has nothing to do with the P-value. We cannot prove the null hypothesis (that the new formula and the old formula are equally effective), but can only fail to find evidence that would cause us to reject it. All we can say about this P-value is that there is a 27% chance of seeing the observed effectiveness from natural sampling variation if the new formula and the old one are equally effective.

31. False claims?

a) Using the normal model, calculate the z-value: $z = \dfrac{(\hat{p} - p_0)}{SD(\hat{p})}$; $SD(\hat{p}) = \sqrt{\dfrac{p_0 q_0}{n}} = \sqrt{\dfrac{0.5 * 0.5}{20}} = 0.112$

$$z = \frac{(0.60 - 0.50)}{0.112} = 0.893 \text{ resulting in a P-value} = 0.186.$$

Using exact probabilities, this is a binomial probability of P(12 or more red out of 20 when p = 0.5). Using technology $P(X \geq 12) = 0.252$. Or using a normal approximation for the binomial, this is the $P(X \geq 11.5$, when $\mu = .5(20) = 10$, and $\sigma = \sqrt{(20)(0.50)(0.50)} = 2.236)$. Using technology here, the approximate probability is 0.251.

b) A hypothesis test:
  H_0: $p = 0.50$
  H_A: $p \neq 0.50$

$$z = \frac{(\hat{p} - p_0)}{SD(\hat{p})}; \quad SD(\hat{p}) = \sqrt{\frac{p_0 q_0}{n}} = \sqrt{\frac{0.5 * 0.5}{20}} = 0.112$$

$$z = \frac{(0.60 - 0.50)}{0.112} = 0.893 \text{ resulting in a P-value} = 0.186 * 2 = 0.372.$$

There is not strong evidence to conclude that the proportion of reds and greens differ from 0.50. It seems reasonable to think that there really may have been half red and half green. We would expect to get 12 or more reds out of 20 more than 18% of the time, so there's no real evidence that the company's claim is not true.

33. E-commerce. The 1% P-value means that, if the mean monthly sales due to online purchases has not changed, there is a 1% chance (or 1 out of every 100 samples) that the resulting mean sales would occur assuming the historical mean for sales. This is rare and considered to be a significant value.

35. Social security payments.

a) The confidence interval equation is: $\bar{y} \pm t^*_{n-1} X SE(\bar{y})$ where $SE = \dfrac{s}{\sqrt{n}}$. Substituting values:

$$\$915 \pm 1.984 * \frac{90}{\sqrt{100}} = \$915 \pm 17.856 = (\$897.14, \$932.86) \text{ or } \$897.14 \text{ to } \$932.86. \text{ We are 95\%}$$

confident that the interval $897.14 to $932.86 contains the true mean Social Security benefit for widows and widowers in the Texas county.

b) With a P-value of 0.007, the hypothesis test results are significant and we reject the null hypothesis. We conclude that the mean benefit payment for the Texas county is different from the 2005 average monthly benefit for the state of $940. The 95% confidence interval estimate is $897.14 to $932.86. This interval does not contain the hypothesized value of $940 and, therefore, we have evidence that the mean is unlikely to be $940.

37. Spike poll.
a) Conditions are satisfied: random sample; less than 10% of population; more than 10 success and failures; the 98% confidence interval:

$$\hat{p}\pm z*\sqrt{\frac{\hat{p}\hat{q}}{n}}=0.02995\pm2.326\sqrt{\frac{(0.03)(0.97)}{(1302)}}=0.03\pm0.011=(0.019,0.041)=(1.9\%,4.1\%)$$

b) Since 5% is not in the interval, there is strong evidence that fewer than 5% of all men use work as their primary measure of success.

c) The significance level is $\alpha=0.01$; it is a lower tail test based on a 98% confidence interval.

39. Economy.
   a) Conditions are satisfied: a random sample; less than 10% of population; more than 10 success and failures; the 95% confidence interval:

$$\hat{p}\pm z*\sqrt{\frac{\hat{p}\hat{q}}{n}}=0.24\pm1.96\sqrt{\frac{(0.24)(0.76)}{(2336)}}=0.24\pm0.017=(0.223,0.257)=(22.3\%,25.7\%)$$

We are 95% confident that the true proportion of U.S. adults who rate the economy as Excellent/Good is between 0.223 and 0.257.

   b) There is evidence that these reports are significantly different because 0.28 is not within the confidence interval.

   c) The significance level is $\alpha=0.05$; it is a two-tail test based on a 95% confidence interval.

41. Convenient alpha.
   a) Less likely because 5% is less than 7.2%. Lowering α decreases the chance of rejecting the null hypothesis.
   b) Alpha levels must be chosen *before* examining the data. Otherwise, the alpha level could always be selected at a value that would reject the null hypothesis.

43. Product testing.
   1) Null and alternative hypotheses should involve p, not $\hat{p}$.

   2) The question is about *failing* to meet the goal. H_A should be $p<0.96$.

   3) The student failed to check $nq=(200)(0.04)=8$. Since $nq<10$, the Success/Failure condition is violated. Similarly, the 10% Condition is not verified.

   4) $SD(\hat{p})=\sqrt{\frac{pq}{n}}=\sqrt{\frac{(0.96)(0.04)}{200}}\approx0.014.$ The student used $\hat{p}$ and $\hat{q}$.

   5) Value of z is incorrect. The correct value is $z=\dfrac{0.94-0.96}{0.014}\approx-1.43.$

   6) *P*-value is incorrect. $P=P(z<-1.43)=0.076$

   7) For the *P*-value given, an incorrect conclusion is drawn. A *P*-value of 0.12 provides no evidence that the new system has failed to meet the goal. The correct conclusion for the corrected *P*-value is: Since the *P*-value of 0.076 is fairly low, there is weak evidence that the new system has failed to meet the goal.

45. E-commerce, part 2.
   a) The hypotheses are: $H_0:\mu=23.3\ years;H_A:\mu>23.3\ years$.
   b) The **randomization condition**: The 40 online shoppers were selected randomly. **Nearly normal** condition: the distribution of the sample should be examined to check for serious skewness or outliers but the sample of 40 shoppers is large enough that it should be safe to proceed.
   c) Calculate the t value by using the equation: $t_{n-1}=\dfrac{(\bar{y}-\mu_0)}{SE(\bar{y})}$ where $SE(\bar{y})=\dfrac{s}{\sqrt{n}}=\dfrac{5.3}{\sqrt{40}}=0.838$.

Substituting: $t_{n-1} = \dfrac{(\bar{y} - \mu_0)}{SE(\bar{y})} = \dfrac{(24.2 - 23.3)}{0.838} = 1.074$. Calculate the t test and the resulting P-value

using technology: $t_{39}(23.3, \dfrac{s}{\sqrt{n}}) = t_{39}(23.3, 0.838)$ results in a P-value of 0.145.

d) If the mean age of shoppers remains at 23.3 years, there is a 14.5% chance of getting a sample mean of 24.2 years or older simply from natural sampling variation.

e) There is not enough evidence to suggest that the mean age of online shoppers has increased from the mean of 23.3 years.

47. Pricing for competitiveness.
a) In order to determine whether they should market the wax, a t test needs to be conducted to determine whether the champion's times are less than 55 seconds. The hypotheses are:
$H_0 : \mu = 55\,\text{sec}; H_A : \mu < 55\,\text{sec}$. Next, determine whether the necessary assumptions to perform inference are satisfied. The **independence assumption**: the times were selected randomly and we can assume that the times are independent and representative of all champion's times. **Nearly normal** condition: the histogram of the times is unimodal and roughly symmetric.

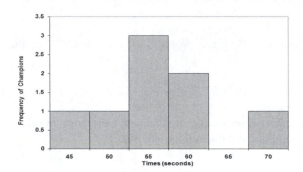

Calculate the t value by using the equation: $t_{n-1} = \dfrac{(\bar{y} - \mu_0)}{SE(\bar{y})}$ where $SE(\bar{y}) = \dfrac{s}{\sqrt{n}} = \dfrac{7.029}{\sqrt{8}} = 2.485$.

Substituting: $t_{n-1} = \dfrac{(\bar{y} - \mu_0)}{SE(\bar{y})} = \dfrac{(53.1 - 55)}{2.485} = -0.765$. Calculate the t test and the resulting P-value using

technology: $t_7(53.1, \dfrac{s}{\sqrt{n}}) = t_7(25.02, 2.485)$ results in a P-value of 0.235. This value fails to reject H_0.

There is little evidence to suggest that the mean time is less than 55 seconds. They should not market the new ski wax.

b) They have made a Type II error. They won't market a competitive wax and thus lose the potential profit from doing so.

49. Environment.
a) Let p be the proportion of children with genetic abnormalities (0.05). This is stated in the null hypothesis. The problem states that chemicals have led to an increase in abnormalities from 5%. This defines the alternative hypothesis as a one-sided test.
$H_0: p = 0.05$
$H_A: p > 0.05$

b) The study doesn't state whether the sample is a SRS so we would have to assume that the sample is representative of the population. 384 children < 10% of all children; $np = (384)(0.05) = 19.2$ which is > 10 and $nq = (384)(0.95) = 364.8 > 10$.

c) Calculate the z-value: $z = \dfrac{(\hat{p} - p_0)}{SD(\hat{p})}$; $SD(\hat{p}) = \sqrt{\dfrac{p_0 q_0}{n}} = \sqrt{\dfrac{0.05 * 0.95}{384}} = 0.0111$; $\hat{p} = 46/384 = 0.12$

$$z = \frac{(0.12 - 0.05)}{0.0111} = 6.31 \text{ resulting in a P-value} < 0.0001 \ (1.4 \times 10^{-10})$$

d) If 5% of children have genetic abnormalities, the chance of observing 46 children with genetic abnormalities in a random sample of 384 children is essentially zero.

e) The conclusion is that we have enough evidence to reject the null hypothesis H_0. There is strong evidence that more than 5% of children have genetic abnormalities.

f) We don't know that environmental chemical cause genetic abnormalities, only that the rate is higher now than in the past. We cannot show causation.

51. Education.

a) Let p be the proportion of students in 2000 with perfect attendance the previous month (0.34). This is stated in the null hypothesis. The problem asks if there is evidence of a decrease in student attendance. This defines the alternative hypothesis as a one sided test (< 34%).

H_0: $p = 0.34$
H_A: $p < 0.34$

b) **Plausible independence condition:** It is reasonable to think that the students' attendance records are independent of one another.
Randomization condition: Although not specifically stated, we can assume that the National Center for Educational Statistics used random sampling.
10% condition: The 8302 students are less than 10% of all students.
Success/Failure condition: $np = (8302)(0.34) = 2822.68$ and $nq = (8302)(0.66) = 5479.32$ are both greater than 10, so the sample is large enough.

c) Calculate the z-value: $z = \dfrac{(\hat{p} - p_0)}{SD(\hat{p})}$; $SD(\hat{p}) = \sqrt{\dfrac{p_0 q_0}{n}} = \sqrt{\dfrac{0.34 * 0.66}{8302}} = 0.0052$

$$z = \frac{(0.33 - 0.34)}{0.0052} = -1.923 \text{ resulting in a P-value} = 0.0272.$$

d) The null hypothesis H_0 is rejected at $\alpha = 0.05$; there is evidence to suggest that the percentage of students with perfect attendance in the previous month has decreased in 2000..

e) This result is statistically significant at $\alpha = 0.05$ not but it is not clear that it has practical significance. The percentage dropped only from 34% to 33%.

53. Retirement.

a) The study doesn't state whether the sample is a SRS so we would have to assume that the sample is representative of the population. $1000 < 10\%$ of all workers; $np = 520$ which is > 10 and $nq = 480$ which is > 10. The 95% confidence interval:

$$\hat{p} \pm z^* \sqrt{\frac{\hat{p}\hat{q}}{n}} = 0.52 \pm 1.96 \sqrt{\frac{(0.52)(0.48)}{(1000)}} = 0.52 \pm 0.031 = (0.489, 0.551) = (48.9\%, 55.1\%)$$

We are 95% confident that between 49.9% and 55.1% of workers have invested in individual retirement accounts.

b) Since 44% is not in the 95% confidence interval, there is strong evidence that the percentage of workers who have invested in individual retirement accounts was not 44%. In fact, our sample indicates an increase in the percentage of adults who invest in individual retirement accounts.

55. Maintenance costs. Let p be the proportion of cars with faulty emissions (0.20). This is stated in the null hypothesis. The problem asks if the proportion is higher than 20%. This defines the alternative hypothesis as a one sided test (> 0.20).

H_0: $p = 0.20$
H_A: $p > 0.20$

Two conditions are not satisfied. 22 is greater than 10% of the population of 150 cars, and $np = (22)(0.20) = 4.4$, which is not greater than 10. It's probably not a good idea to proceed with a hypothesis test.

57. Webzine. Let p be the proportion of readers interested in an online edition (0.25). This is stated in the null hypothesis. The problem wants to go ahead if the percentage is greater than 25%. This defines the alternative hypothesis as a one sided test (> 0.25).

H_0: $p = 0.25$
H_A: $p > 0.25$

Plausible independence condition: Interest of one reader should not affect interest of other readers.

Randomization condition: The magazine conducted an SRS of 500 current readers.

10% condition: 500 readers are less than 10% of all potential subscribers.

Success/Failure condition: $np = (500)(0.25) = 125$ and $nq = (500)(0.75) = 375$ are both greater than 10, so the sample is large enough.

The conditions have been satisfied, so a Normal model can be used to model the sampling distribution of the proportion, with $\mu_{\hat{p}} = p = 0.25$ and $\sigma(\hat{p}) = \sqrt{\dfrac{pq}{n}} = \sqrt{\dfrac{(0.25)(0.75)}{500}} \approx 0.0194$.

We can perform a one-proportion z-test. The observed proportion of interested readers is $\hat{p} = \dfrac{137}{500} = 0.274$

Since the P-value = 0.1076 is high, we fail to reject the null hypothesis. There is little evidence to suggest that the proportion of interested readers is greater than 25%. The magazine should not publish the online edition.

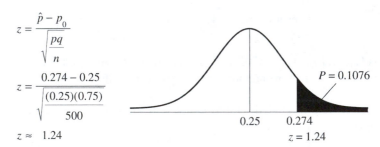

59. Women executives. Let p be the hypothesized proportion of female executives (0.40). This is stated in the null hypothesis. The company wants to see if the sample results are less than 40%. This defines the alternative hypothesis as a one sided test (< 0.40).

H_0: $p = 0.40$
H_A: $p < 0.40$

Plausible independence condition: It is reasonable to think that executives at this company were chosen independently.

Randomization condition: The executives were not chosen randomly, but it is reasonable to think of these executives as representative of all potential executives over many years.

10% condition: The 43 executives represent all executives at this company; consequently, results may not be able to be generalized to all companies.

Success/Failure condition: $np = (43)(0.40) = 17.2$ and $nq = (43)(0.60) = 25.8$ are both greater than 10, so the sample is large enough.

The conditions have been satisfied, so a Normal model can be used to model the sampling distribution of the proportion, with $\mu_{\hat{p}} = p = 0.40$ and $\sigma(\hat{p}) = \sqrt{\dfrac{pq}{n}} = \sqrt{\dfrac{(0.40)(0.60)}{43}} \approx 0.0747$.

We can perform a one-proportion z-test. The observed proportion of female executives is $\hat{p} = \dfrac{13}{43} \approx 0.302$.

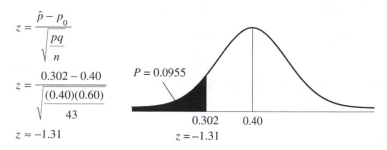

$$z = \dfrac{\hat{p} - p_0}{\sqrt{\dfrac{pq}{n}}}$$

$$z = \dfrac{0.302 - 0.40}{\sqrt{\dfrac{(0.40)(0.60)}{43}}}$$

$$z \approx -1.31$$

$P = 0.0955$

0.302   0.40

$z = -1.31$

Since the P-value = 0.0955 is high, we fail to reject the null hypothesis at $\alpha = 0.05$. There is little evidence to suggest proportion of female executives is any different from the overall proportion of 40% female employees at the company.

61. Nonprofit. Let p be the proportion of high school dropouts for the year 2000 (0.109). This is stated in the null hypothesis. The school district wants to see if tutoring has improved the dropout rate (< 10.9%). This defines the alternative hypothesis as a one sided test (< 0.109).

   H_0: $p = 0.109$
   H_A: $p < 0.109$

Plausible independence condition/Randomization condition: Assume that the students at this high school are representative of all students nationally. This is really what we are testing. The dropout rate at this high school has traditionally been close to the national rate. If we reject the null hypothesis, we will have evidence that the dropout rate at this high school is no longer close to the national rate.
 10% condition: 1782 students are less than 10% of all students nationally.
Success/Failure condition: $np = (1782)(0.109) = 194.238$ and $nq = (1782)(0.891) = 1587.762$ are both greater than 10, so the sample is large enough.

The conditions have been satisfied, so a Normal model can be used to model the sampling distribution of the proportion, $\mu_{\hat{p}} = p = 0.103$ and $\sigma(\hat{p}) = \sqrt{\dfrac{pq}{n}} = \sqrt{\dfrac{(0.103)(0.897)}{1782}} \approx 0.0072$.

We can perform a one-proportion z-test. The observed proportion of dropouts is

$\hat{p} = \dfrac{175}{1782} = 0.0982$; $z = \dfrac{\hat{p} - p_0}{\sqrt{\dfrac{pq}{n}}} = \dfrac{0.0982 - 0.109}{\sqrt{\dfrac{(0.109)(0.891)}{1782}}} = -1.463$ resulting in a P-value of 0.072.

Since the P-value is above 5%, there is insufficient evidence at $\alpha = 0.05$ to determine that there is a decrease in the dropout rate from 10.9%.

63. Public relations. Let p be the proportion of lost luggage returned the next day (0.90). This is stated in the null hypothesis. A consumer group wants to investigate to see if the proportion is less than 90% (< 90%). This defines the alternative hypothesis as a one sided test (< 0.90).

   H_0: $p = 0.90$
   H_A: $p < 0.90$

Plausible independence condition: It is reasonable to think that the people surveyed were independent with regards to their luggage woes.
Randomization condition: Although not stated, we will hope that the survey was conducted randomly, or at least that these air travelers are representative of all air travelers for that airline.

10% condition: 122 air travelers are less than 10% of all air travelers on the airline.
Success/Failure condition: $np = (122)(0.90) = 109.8$ and $nq = (122)(0.10) = 12.2$ are both greater than 10, so the sample is large enough.

The conditions have been satisfied, so a Normal model can be used to model the sampling distribution of the proportion, with $\mu_{\hat{p}} = p = 0.90$ and $\sigma(\hat{p}) = \sqrt{\dfrac{pq}{n}} = \sqrt{\dfrac{(0.90)(0.10)}{122}} \approx 0.0272$.

We can perform a one-proportion z-test. The observed proportion of dropouts is $\hat{p} = \dfrac{103}{122} \approx 0.844$.

Since the P-value = 0.0201 is low, we reject the null hypothesis. There is evidence that the proportion of lost luggage returned the next day is lower than the 90% claimed by the airline.

$$z = \frac{\hat{p} - p_0}{\sqrt{\dfrac{pq}{n}}}$$

$$z = \frac{0.844 - 0.90}{\sqrt{\dfrac{(0.90)(0.10)}{122}}}$$

$$z \approx -2.05$$

$P = 0.0201$

0.844   0.90
$z = -2.05$

65. **Business ethics.** Let p be the proportion of newly hired MBA students confronted with unethical practices during their first year of employment (0.30). This is stated in the null hypothesis. The dean wants to know if her school's MBA program is similar ($\neq 30\%$). This defines the alternative hypothesis as a one sided test ($\neq 0.30$).

H_0: $p = 0.30$
H_A: $p \neq 0.30$

Plausible independence condition/Randomization condition: There is no reason to believe that students' rates would influence others; the professor considers this class typical of other classes.
10% condition: 120 graduates are less than 10% of all students in the MBA program.
Success/Failure condition: $np = (120)(0.30) = 36$ and $nq = (120)(0.70) = 84$ are both > 10, so the sample is large enough.
The conditions have been satisfied, so we can perform a one-proportion z-test.
The observed proportion of first-time buyers making below $40,000 is $\hat{p} = 0.27$.

Calculate the z-value: $z = \dfrac{(\hat{p} - p_0)}{SD(\hat{p})}$; $SD(\hat{p}) = \sqrt{\dfrac{p_0 q_0}{n}} = \sqrt{\dfrac{0.30 * 0.70}{120}} = 0.04183$

$z = \dfrac{(0.27 - 0.30)}{0.04183} = -0.717$ resulting in a P-value = 0.4734.

Since the P-value is > 0.05, we fail to reject the null hypothesis H_0 at $\alpha = 0.05$. There is not convincing evidence that the rate at which these students are exposed to unethical business practices is different from that reported in the study.

67. **U.S. politics.**
   a) The observed proportion of stocks increasing in value $\hat{p} = 0.71$.

   $z = \dfrac{(\hat{p} - p_0)}{SD(\hat{p})}$; $SD(\hat{p}) = \sqrt{\dfrac{p_0 q_0}{n}} = \sqrt{\dfrac{0.58 * 0.42}{2020}} = 0.011$ $z = \dfrac{(0.71 - 0.58)}{0.011} = 11.8$

   b) For a two-sided significance alternative hypothesis using $\alpha = 0.1\% = 0.001$; The critical z-value for 0.001/2 or 0.0005 is 3.29.

    c)   We conclude that the percent of U.S. adults giving "quite a lot" of thought to the upcoming election is significantly different in 2008 than it was in 2004.

69. Testing cars.

    a)   In this context, a Type I error is concluding that the shop is not meeting standards when it actually is.

    b)   In this context, a Type II error is when the regulators certify the shop when it is not meeting the standards.

    c)   The shop's owner would consider a Type I error to be more serious because that error would state that the shop is not meeting standards when it actually is and would affect business.

    d)   Environmentalists would consider a Type II error to be more serious because that error would state that the shop is meeting standards when it actually is not and could be polluting.

71. Testing cars, again.

    a)   In this context, the power of the test is the probability of detecting that the shop is not meeting standards when they are not.

    b)   40 cars produces a higher power value because n is larger. For the same significance level, an increase in the sample size results in a decrease in the probability of a Type II error, resulting in an increase in the power of the test.

    c)   The power will be greater if they use a 10% level of significance because there will be a greater chance to reject the null hypothesis.

    d)   The power will be greater if the repair shop's inspectors are a lot out of compliance; larger problems are easier to detect.

73. Statistics software.

    a)   One-tailed; we are testing to see if a decrease in the dropout rate is associated with the software.

    b) H_0: p = 0.13 (the dropout rate does not change following the use of the software)
        H_A: p < 0.13 (the dropout rate decreases following the use of the software)

    c)   The professor buys the software when the dropout rate has not actually decreased (the null hypothesis is true, but we mistakenly reject it).

    d)   The professor doesn't buy the software when the dropout rate has actually decreased (the null hypothesis is false, but he fails to reject it).

    e)   The power of the test is the probability of buying the software when the dropout rate has actually decreased (the power of a test is the probability that it correctly rejects a false null hypothesis).

75. Statistics software, part 2.

    a) H_0: $p = 0.13$ (the dropout rate does not change following the use of the software)
        H_A: $p < 0.13$ (the dropout rate decrease following the use of the software)
    Plausible independence condition/Randomization condition: There is no reason to believe that one student dropping has an influence on another student dropping.
    This year's class of 203 can be considered representative of all statistics students.
    10% condition: 203 < 10% of all students.
    Success/Failure condition: $np = (203)(0.13) = 26.39 > 10$ and $nq = (203)(0.87) = 176.61 > 10$ so the sample is large enough.
    The conditions have been satisfied, so we can perform a one-proportion z-test.
    The observed proportion of students dropping statistics $\hat{p} = 11/203 = 0.0542$.

Calculate the z-value: $z = \dfrac{(\hat{p} - p_0)}{SD(\hat{p})}$; $SD(\hat{p}) = \sqrt{\dfrac{p_0 q_0}{n}} = \sqrt{\dfrac{0.13 * 0.87}{203}} = 0.0236$

$z = \dfrac{(0.0542 - 0.13)}{0.0236} = -3.21$ resulting in a P-value = 0.0007. Since the P-value is very low, we reject the null hypothesis H_0. There is strong evidence that the dropout rate has dropped since the use of the software program was implemented. As long as the professor feels confident that this class of students is representative of all potential statistics students, then he should buy the program and continue its use.

b)   The chance of observing 11 or fewer dropouts in a class of 203 is only 0.07% if the dropout rate is really 13%.

Chapter 14 – Comparing Two Groups

SECTION 14.1

1.

a) The sample mean ages for each neighborhood are $\bar{y}_1 = 58.0\,years$ and $\bar{y}_2 = 47.0\,years$.

b) $\bar{y}_1 - \bar{y}_2 = 11\,years$ is the estimated difference of the mean ages of the two neighborhoods.

c) The sample variances in ages for each neighborhood are

$$s_1^2 = \frac{\sum(y - \bar{y}_1)^2}{n_1 - 1} = \frac{(57-58)^2 + (60-58)^2 + (46-58)^2 + (62-58)^2 + (67-58)^2 + (56-58)^2}{6-1} = 50\,years^2$$

$$s_2^2 = \frac{\sum(y - \bar{y}_2)^2}{n_2 - 1} = \frac{(50-47)^2 + (43-47)^2 + (35-47)^2 + (53-47)^2 + (46-47)^2 + (55-47)^2}{6-1} = 54\,years^2$$

d) The sample standard deviation in ages for each neighborhood are

$$s_1 = \sqrt{s_1^2} = \sqrt{50\,years^2} = 7.07\,years$$

$$s_2 = \sqrt{s_2^2} = \sqrt{54\,years^2} = 7.35\,years$$

e) The standard error of the difference of the two sample means is

$$SE(\bar{y}_1 - \bar{y}_2) = \sqrt{\frac{s_1^2}{n_1} + \frac{s_2^2}{n_2}} = \sqrt{\frac{50}{6} + \frac{54}{6}} = 4.16\,years$$

3.

a) $\bar{y}_1 - \bar{y}_2 = 57.2 - 47.6 = 9.6\,years$ is the estimated difference of the mean ages between the two neighborhoods.

b) The standard error of the estimated mean difference is

$$SE(\bar{y}_1 - \bar{y}_2) = \sqrt{\frac{s_1^2}{n_1} + \frac{s_2^2}{n_2}} = \sqrt{\frac{(7.51)^2}{30} + \frac{(7.85)^2}{35}} = 1.91\,years$$

c) The *t*-statistic for the observed difference (assuming that the true mean difference is 0) is

$$t = \frac{(\bar{y}_1 - \bar{y}_2) - (\mu_1 - \mu_2)}{SE(\bar{y}_1 - \bar{y}_2)} = \frac{9.6}{1.91} = 5.03$$

SECTION 14.2

5.

    a)   The t-statistic for the observed difference (assuming that the true mean difference is 0) is

$$t = \frac{(\bar{y}_1 - \bar{y}_2) - (\mu_1 - \mu_2)}{SE(\bar{y}_1 - \bar{y}_2)} = \frac{11}{4.16} = 2.64$$

    b)   The degrees of freedom are

$$df = \frac{\left(\dfrac{s_1^2}{n_1} + \dfrac{s_2^2}{n_2}\right)^2}{\dfrac{1}{n_1-1}\left(\dfrac{s_1^2}{n_1}\right)^2 + \dfrac{1}{n_2-1}\left(\dfrac{s_2^2}{n_2}\right)^2} = \frac{\left(\dfrac{50}{6} + \dfrac{54}{6}\right)^2}{\dfrac{1}{5}\left(\dfrac{50}{6}\right)^2 + \dfrac{1}{5}\left(\dfrac{54}{6}\right)^2} = 9.98$$

    c)   The degrees of freedom are 5.
    d)   Using technology the P-value is 0.0247. If we use 10 df and look in Table T, we can say that the P-value is between 0.02 and 0.05.
    e)   Using technology the P-value is 0.0459. If we use 5 df and look in Table T, we can say that the P-value is between 0.02 and 0.05.
    f)   For either method of calculating df, the P-value is less than $\alpha=0.05$. Therefore we can reject the null hypothesis and conclude that there is strong evidence to suggest that the mean age of houses is not the same in the two neighborhoods.

7.

    a)   Using technology, the P-value < 0.001.
    b)   The $df = 29$ using this rule. Using technology, the P-value < 0.001.
    c)   For either method of calculating df, the P-value is less than $\alpha=0.05$. Therefore we can reject the null hypothesis and conclude that there is strong evidence to suggest that the mean age of houses is not the same in the two neighborhoods.

SECTION 14.4

9.

    a)   The 95% confidence interval for the mean difference in ages of houses in the two neighborhoods (using $df = 10$) is

$$(\bar{y}_1 - \bar{y}_2) \pm t^*_{df} SE(\bar{y}_1 - \bar{y}_2)$$
$$(58 - 47) \pm 2.228(4.16) = (1.72, 20.28) \text{ years.}$$

    b)   Zero is not contained within the confidence interval.
    c)   It suggests that we should reject the null hypothesis since 0 is not a plausible value for the true mean difference.

11.

    a)   The 95% confidence interval for the mean difference in ages of houses in the two neighborhoods ($df = 53.4$; for Table T round and use $df = 50$) is

$$(\bar{y}_1 - \bar{y}_2) \pm t^*_{df} SE(\bar{y}_1 - \bar{y}_2)$$

$(57.2 - 47.6) \pm 2.009(1.91) = (5.77, 13.43)$ years.

b) This confidence interval is narrower because the sample sizes are larger.

c) Zero is not contained within the confidence interval.

d) It suggests that we should reject the null hypothesis since 0 is not a plausible value for the true mean difference.

SECTION 14.5

13.

a) The pooled t-statistic is

$$s^2_{pooled} = \frac{(n_1 - 1)s_1^2 + (n_2 - 1)s_2^2}{(n_1 - 1) + (n_2 - 1)} = \frac{(6 - 1)(50) + (6 - 1)(54)}{(6 - 1) + (6 - 1)} = 52$$

$$SE_{pooled}(\bar{y}_1 - \bar{y}_2) = s_{pooled}\sqrt{\frac{1}{n_1} + \frac{1}{n_2}} = 7.21\sqrt{\frac{1}{6} + \frac{1}{6}} = 4.16$$

$$t = \frac{(\bar{y}_1 - \bar{y}_2) - 0}{SE_{pooled}(\bar{y}_1 - \bar{y}_2)} = \frac{(58 - 47) - 0}{4.16} = 2.64$$

The P-value is 0.0246.
Conclusion: Reject the null hypothesis; there is sufficient evidence to suggest that the true means are different.

b) The 95% confidence interval (using 10 df) is

$$(\bar{y}_1 - \bar{y}_2) \pm t^*_{df} \times SE_{pooled}(\bar{y}_1 - \bar{y}_2)$$

$(58 - 47) \pm 2.228 \times 4.16$

$(1.72, 20.28)$ years

c) Not different. The two methods will result in essentially the same confidence intervals and hypothesis test results because the standard errors are the same.

15.

a) The pooled t-statistic is

$$s^2_{pooled} = \frac{(n_1 - 1)s_1^2 + (n_2 - 1)s_2^2}{(n_1 - 1) + (n_2 - 1)} = \frac{(30 - 1)(7.51)^2 + (35 - 1)(7.85)^2}{(30 - 1) + (35 - 1)} = 59.218$$

$$SE_{pooled}(\bar{y}_1 - \bar{y}_2) = s_{pooled}\sqrt{\frac{1}{n_1} + \frac{1}{n_2}} = 7.695\sqrt{\frac{1}{30} + \frac{1}{35}} = 1.914$$

$$t = \frac{(\bar{y}_1 - \bar{y}_2) - 0}{SE_{pooled}(\bar{y}_1 - \bar{y}_2)} = \frac{(57.2 - 47.6) - 0}{1.914} = 5.01$$

The P-value is < 0.001.
Conclusion: Reject the null hypothesis; there is sufficient evidence to suggest that the true means are different.

b)   The 95% confidence interval ($df = 63$; in Table T use 60 df) is

$$(\bar{y}_1 - \bar{y}_2) \pm t^*_{df} \times SE_{pooled}(\bar{y}_1 - \bar{y}_2)$$

$$(57.2 - 47.6) \pm 2.000 \times 1.914$$

$$(5.77, 13.43) \text{ years}$$

c)   The results are essentially the same. Because the standard errors are so close in value, the two methods will result in essentially the same confidence intervals and hypothesis test results.

SECTION 14.6

17.

Tukey's test is applicable because the groups are independent and one group (Neighborhood 2) has the lowest age value (35) and the other group (Neighborhood 1) has the highest age value (67). To perform Tukey's test, we see that there are 2 observations in Neighborhood 2 that are lower than the lowest value in Neighborhood 1 and that there are 5 observations in Neighborhood 1 that are higher than the highest value in Neighborhood 2. This gives us a total of 7. The difference is significant at $\alpha = 0.05$.

SECTION 14.7

19.

a)   The data are paired. The same volunteer participating in the study used both chairs.
b)   The data should be treated as independent. The samples are selected randomly from two different neighborhoods.
c)   The data are paired. The productivity of the two workers is measured for the same hours.

SECTION 14.8

21.

a)   The data are paired. Customer traffic is recorded for each of 10 stores with and without the "buy one, get one free" campaign. The data are collected on the same sample of 10 stores, therefore they are paired by store.
b)   The sample of differences is 4, -2, 2, 7, 4, 0, 7, -6, 8, and 6. The mean of these differences is 3 customers.
c)   The standard deviation of these differences is

$$s_d = \sqrt{\frac{(4-3)^2 + (-2-3)^2 + (2-3)^2 + (7-3)^2 + (4-3)^2 + (0-3)^2 + (7-3)^2 + (-6-3)^2 + (8-3)^2 + (6-3)^2}{10-1}}$$

$$s_d = 4.52 \text{ customers}$$

d)   The standard error of the mean difference is

$$SE(\bar{d}) = \frac{s_d}{\sqrt{n}} = \frac{4.52}{\sqrt{10}} = 1.43 \text{ customers}$$

e)   The value of the t statistic is

$$t = \frac{\bar{d} - \Delta_0}{SE(\bar{d})} = \frac{3 - 0}{1.43} = 2.098$$

f)   The *t* statistic has *n-1* or 9 *df*.

g)   Since the supermarket wants to know if the campaign increases store traffic, the test is one-sided.

h)   The P-value is 0.0327.

i)   Since the P-value is less than $\alpha = 0.05$, we can reject the null hypothesis and conclude that the campaign has increased store traffic.

23.

The 90% confidence interval (using 9 *df*) for the mean difference is

$$\bar{d} \pm t^*_{n-1} \times SE(\bar{d})$$
$$3 \pm 1.833 \times 1.43$$
$$(0.379, 5.621) \text{ customers.}$$

CHAPTER EXERCISES

25. Hot dogs and calories. The P-value of 0.124 is > 0.10 (corresponding to, which can be considered to be a reasonable α) and is too large a value to reject H_0. The data does not support that there is a difference in mean calorie content between meat hot dogs and beef hot dogs.

27. Learning math.

a)   The margin of error is half of the interval given. $11.427 - 5.573 = 5.854 \div 2 = 2.927$.

b)   A 98% confidence interval requires a larger margin of error in order to be more confident. Mathematically, the critical *t* value is larger for a higher confidence level making the margin of error larger.

c)   We are 95% confident that the mean score for the CPMP math students will be between 5.573 and 11.427 points higher on this assessment than the mean score of the traditional students.

d)   Because the entire interval is above zero, there is strong evidence that students who learn with CPMP will have higher mean scores in applied algebra than those in traditional programs.

29. CPMP, again.

a)   The hypotheses are: $H_0 : \mu_C - \mu_T = 0$; $H_A : \mu_C - \mu_T \neq 0$

b)   If the null hypothesis is true, that is, if the mean scores for the CPMP and traditional students are really equal, there is less than a 1 in 10,000 chance of seeing a difference as large as or larger than the observed difference of 9.4 points (=38.4-29.0) resulting only from natural sampling variation.

c)   There is strong evidence that the CPMP students have a different mean score than the traditional students. The evidence suggests that the CPMP students have a higher mean score.

31. CPMP and word problems. The hypotheses are: $H_0 : \mu_C - \mu_T = 0$; $H_A : \mu_C - \mu_T \neq 0$. Independent group assumption: scores of students from different classes can be assumed to be independent from each other. Randomization condition: it is not specifically stated in the problem whether or not the classes were randomly assigned. However, the study could be looked up in the journal identified in Exercise 3. Being a scholarly journal, it is likely that the classes were randomly assigned to either CPMP or traditional curricula. 10% condition: 320 and 273 are less than 10% of all students. Nearly Normal condition: we don't have access to the actual data so we are unable to check the distribution of the sample. However, the sample sizes are large and the Central Limit theorem allows us to proceed.

The P-value is obtained through technology methods: The $t^* = 1.406$ for $df = 590.05$. The resulting P-value for a 2-sided test is 0.1602 or over 16%. This value is > 5% and fails to reject the null hypothesis. There is not sufficient evidence to believe that the CPMP students have a different mean score on the word problems test than the traditional students.

33. Trucking company.

   a) The confidence interval is: $(\bar{y}_1 - \bar{y}_2) \pm t^*_{df} \times \mathrm{SE}(\bar{y}_1 - \bar{y}_2)$ where the standard error of the difference

   of the means is: $\mathrm{SE}(\bar{y}_B - \bar{y}_A) = \sqrt{\dfrac{s_1^2}{n_1} + \dfrac{s_2^2}{n_2}}$ Calculate the observed difference in the two means:

   $\bar{y}_B - \bar{y}_A = 43 - 40 = 3$ (it is usually easier to work with a positive difference).

   $\mathrm{SE}(\bar{y}_B - \bar{y}_A) = \sqrt{\dfrac{(3)^2}{20} + \dfrac{(2)^2}{20}} = 0.806$. From technology, the degrees of freedom are 33.1. The

   critical t value for df of 33.1 (manually calculate or from technology) for a confidence level of 95% is 2.0345. The 95% confidence interval is: $3 \pm 2.0345 * 0.806 = 3 \pm 1.64 = (1.36, 4.64)$.

   b) The confidence interval does not contain the value zero (no difference value). Therefore, there is evidence that Route A is faster on average.

35. Cereal company.

   a) The hypotheses are: $H_0 : \mu_C - \mu_A = 0; H_A : \mu_C - \mu_A \neq 0$.

   b) Independent group assumption: the percentage of sugar in the children's cereals is unrelated to the percentage of sugar in adult cereals. Randomization condition: it is reasonable to assume that the cereals are representative of all children's cereals and adult cereals in regard to sugar content. Nearly Normal condition: the histogram of adult cereal sugar content is skewed to the right, but the sample sizes are of reasonable size. The Central Limit Theorem allows us to proceed.

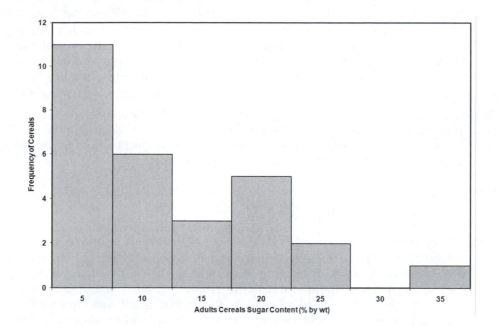

c)   The confidence interval is $(\bar{y}_1 - \bar{y}_2) \pm t^*_{df} \times SE(\bar{y}_1 - \bar{y}_2)$ where the standard error of the difference of

the means is: $SE(\bar{y}_C - \bar{y}_A) = \sqrt{\dfrac{s_1^2}{n_1} + \dfrac{s_2^2}{n_2}}$ Calculate the observed difference in the two means:

$\bar{y}_C - \bar{y}_A = 46.8 - 10.1536 = 36.6464$ (it is usually easier to work with a positive difference).

$SE(\bar{y}_C - \bar{y}_A) = \sqrt{\dfrac{(6.418)^2}{19} + \dfrac{(7.6124)^2}{28}} = 2.0585$. From technology, the degrees of freedom are
42.78. The critical t value for df of 42.78 is 2.018 (manually calculate or from technology). The
95% confidence interval is: $36.65 \pm 2.018 * 2.0585 = 36.65 \pm 4.154 = (32.49, 40.80)$.

d)   The 95% confidence interval does not contain zero (no difference value) so for a two-sided
hypothesis, we can conclude that the mean sugar content for the two cereals is significantly
different at the 5% level of significance.

37. Investment.

a)   The hypotheses are: $H_0 : \mu_C - \mu_D = 0; H_A : \mu_C - \mu_D \neq 0$.

b)   The confidence interval is: $(\bar{y}_1 - \bar{y}_2) \pm t^*_{df} \times SE(\bar{y}_1 - \bar{y}_2)$ where the standard error of the difference

of the means is: $SE(\bar{y}_C - \bar{y}_D) = \sqrt{\dfrac{s_1^2}{n_1} + \dfrac{s_2^2}{n_2}}$ Calculate the observed difference in the two means:

$\bar{y}_C - \bar{y}_D = 9.382 - 8.563 = 0.819$. $SE(\bar{y}_C - \bar{y}_D) = \sqrt{\dfrac{(2.675)^2}{66} + \dfrac{(3.719)^2}{74}} = 0.5434$. From

technology, the degrees of freedom are 132.3. The critical t value for df of 132.3 is 1.978
(manually calculate or from technology). The 95% confidence interval is:
$0.819 \pm 1.978 * 0.5434 = 0.819 \pm 1.075 = (-0.256, 1.894)$.

c)   Because the confidence interval contains zero (no difference value), there is no evidence that the
mean return over a 5-year period is different for consistent style funds as opposed to style drifters.

39. Product testing. The hypotheses are: $H_0 : \mu_N - \mu_C = 0$; $H_A : \mu_N - \mu_C > 0$. Independent groups assumption: student scores in one group should not have an impact on the scores of students in the other group. Randomization condition: students were randomly assigned to classes. Nearly Normal condition: the stemplots of the scores indicate that the distributions are unimodal and approximately symmetric. The question of significance can be answered by calculating the P-value using technology.

We calculate: $\bar{y}_{new} = 51.722$, $\bar{y}_{control} = 41.818$, $s_{new} = 11.706$. $s_{control} = 16.598$, yielding $t = 2.207$. The P-value is obtained through technology methods: The P-value for a one-sided test is 0.0168 which is < 0.05. Therefore, the result is significant and we have evidence to reject the null hypothesis. There is evidence that the students taught using the new activities have a higher mean score on the reading comprehension test than the students taught using traditional methods.

41. Acid streams.

    a) The hypotheses are: $H_0 : \mu_L - \mu_S = 0$; $H_A : \mu_L - \mu_S \neq 0$.

    b) Independent groups assumption: pH levels from the two types of streams are independent. We don't know if the streams were chosen randomly so we would have to assume that the pH level of one stream does not affect the pH of another stream. This seems reasonable. Nearly Normal condition: the boxplots provided show that the pH levels of the streams may be skewed (the limestone distribution has a long whisker up to the first quartile and a low outlier indicating a left skew; the mixed distribution has only the upper quartile whisker indicating a right skew; the shale distribution has the median aligned with either the first or third quartile with a high outlier indicating a possible right skew), however, there are 133 degrees of freedom so we know that the sample sizes are large enough for t-testing. It should be safe to proceed.

    c) The results given show that the P-value is < 0.0001 which is significant and we can reject the null hypothesis. There is strong evidence that the streams with limestone substrates have pH levels different than those of streams with shale substrates. The limestone streams are shown to be less acidic on average.

43. Ginko test.

    a) If the mean memory scores for people taking gingko biloba and people not taking it are the same (null hypothesis), there is a 93.74% chance of seeing a difference in mean memory score this large or larger simply from natural sampling variability.

    b) The P-value is high at 93.74%, therefore, this is a very high percentage chance that a difference in mean memory score this large or larger is due to natural sampling variability.

    c) If proponents of ginkgo biloba continue to insist that it works, they are saying that the analyzers are accepting the null hypothesis when they should be accepting the alternative and rejecting the null hypothesis. This is a Type II error.

45. Productivity.

    a) The confidence interval equation is: $\bar{y} \pm t^{*}_{n-1} \times SE(\bar{y})$ where $SE = \dfrac{s}{\sqrt{n}}$. The critical t value can be for a confidence level of 95% at degrees of freedom (df) of 49 (n-1) is 2.01. For the males, substituting into the equation $SE = \dfrac{s}{\sqrt{n}} = \dfrac{2.52}{\sqrt{50}} = 0.356$, the 95% CI for the male applicants becomes $19.39 \pm 2.01*0.356 = 19.39 \pm 0.72 = (18.67, 20.11)$ pegs. For the females, the corresponding 95% CI for the female applicants has a standard error of $SE = \dfrac{s}{\sqrt{n}} = \dfrac{3.39}{\sqrt{50}} = 0.479$ yielding $17.91 \pm 2.01*0.479 = 17.91 \pm 0.96 = (16.95, 18.87)$ pegs.

b) No conclusions can be drawn from comparing the individual confidence intervals. Looking at the male and female confidence intervals, it appears that the results overlap. However, in order to investigate the difference in male and female means, a two sample t-interval needs to be calculated.

c) The confidence interval is: $(\bar{y}_1 - \bar{y}_2) \pm t_{df}^* \times SE(\bar{y}_1 - \bar{y}_2)$ where the standard error of the difference of the means is: $SE(\bar{y}_M - \bar{y}_F) = \sqrt{\dfrac{s_1^2}{n_1} + \dfrac{s_2^2}{n_2}}$ Calculate the observed difference in the two means:

$\bar{y}_M - \bar{y}_F = 19.39 - 17.91 = 1.48$. $SE(\bar{y}_M - \bar{y}_F) = \sqrt{\dfrac{(2.52)^2}{50} + \dfrac{(3.39)^2}{50}} = 0.597$. From technology, the degrees of freedom are 90.5. The critical t value for df of 90.5 is 1.987 (manually calculate or from technology). The 95% confidence interval is:
$1.48 \pm 1.987 * 0.597 = 1.48 \pm 1.186 = (0.29, 2.67)$ pegs.

d) We are 95% confident that the mean number of pegs placed by males is between 0.29 and 2.67 pegs higher than the mean number of pegs placed by females.

e) The correct result is the two sample t-interval.

f) Mathematically, the problem cannot be solved correctly using two separate confidence intervals because you are adding standing deviations, rather than variances. The two sample difference of means method takes this into account.

47. Designated hitter 2009, part 2.

a) The hypotheses are: $H_0 : \mu_A - \mu_N = 0; H_A : \mu_A - \mu_N > 0$.

b) The confidence interval is: $(\bar{y}_1 - \bar{y}_2) \pm t_{df}^* \times SE(\bar{y}_1 - \bar{y}_2)$ where the standard error of the difference of the means is: $SE(\bar{y}_M - \bar{y}_F) = \sqrt{\dfrac{s_1^2}{n_1} + \dfrac{s_2^2}{n_2}}$ Calculate the observed difference in the two means:

$\bar{y}_{AL} - \bar{y}_{NL} = 4.82 - 4.43 = 0.39; SE(\bar{y}_{AL} - \bar{y}_{NL}) = \sqrt{\dfrac{(0.467)^2}{14} + \dfrac{(0.376)^2}{16}} = 0.156$

From technology, the degrees of freedom are 19.5. The critical t value for df of 19.5 is 2.093 (manually calculate or from technology). The 95% confidence interval is:
$0.39 \pm 2.093 \times 0.156 = (0.06, 0.72)$. We are 95% confident that the mean number of runs scored by American League teams is between 0.06 and 0.72 runs higher than the mean number of runs scored by National League teams.

c) The t-value is 2.47 and the resulting P-value is 0.011 which is < 5%. Therefore, the result is significant and we can reject the null hypothesis.

d) There is strong evidence that the mean number of runs scored per game by the American League is greater than the National league. You would observe a difference in means as large as or larger than this only 1.3% of the time if the mean number of runs scored per game were the same for both leagues.

49. Water hardness.

a) The hypotheses are: $H_0 : \mu_N - \mu_S = 0; H_A : \mu_N - \mu_S \neq 0$. The t-value is 6.47 and the P-value is 3.27×10^{-8} which is < 5%. Therefore, the result is significant and we can reject the null hypothesis. There is strong evidence that the mean mortality rate is different for towns north and south of Derby. There is evidence that the mortality rate north of Derby is higher.

    b)   There is an outlier shown in the boxplot for a data point north of Derby. Therefore, the conditions for inference are not satisfied and it is risky to use the two-sample t-test although the sample size is fairly large. The outlier should be investigated and identified and then the analysis should be repeated with the outlier removed.

51. Job satisfaction.

    a)   The paired t-test is appropriate for these data. The paired data assumption is met as the job satisfaction scores are measured on the same workers before and after the implementation of the exercise break. The randomization condition is satisfied because the 10 workers were randomly selected. A histogram of the differences shows that the distribution is fairly symmetric and unimodal, so the normality assumption appears reasonable.

    b)   Appropriate hypotheses are H_0: $\mu_{After} - \mu_{Before} = 0$; H_A: $\mu_{After} - \mu_{Before} > 0$. Using technology, $t = 3.60$; P-value = 0.003; this value is less than 0.05 therefore we can reject the null hypothesis and conclude that there is sufficient evidence to suggest that the exercise break improved workers' job satisfaction.

53. Delivery time. The hypotheses are: $H_0 : \mu_J - \mu_A = 0$; $H_A : \mu_J - \mu_A \neq 0$. $t = -1.17$; P-value = 0.274; this value is > 5% and we fail to reject the null hypothesis. Although the mean delivery time during August is higher, the difference in delivery time from June is not significant. It may be worthwhile to analyze with a larger sample to see if it produces a different result.

55. Ad campaign.

    a)   We are 95% confident that the mean number of ads remember by viewers of shows with violent content will be between 1.6 and 0.6 lower than the mean number of brand names remembered by viewers of shows with neutral content.

    b)   If they want the viewers to remember their brand names, the company should consider advertising on shows with neutral content as opposed to shows with violent content.

57. Ad recall.

    a)   She might conclude that the mean number of brand names recalled is greater after 24 hours. The P-value is very small, indicating that the mean number of brand names recalled is not the same for the two time periods.

    b)   The groups were not independent. They consist of the same people, asked at two different time periods.

    c)   Because the study was done on the same people, those who had higher recall right after the show might also tend to have higher recall in 24 hours. Also, the first interview may have helped the people remember the brand names for a longer period of time than they would have otherwise.

    d)   Randomly assign half of the group watching a specific type of content to be interviewed immediately after watching and the other half to be interviewed 24 hours later.

59. Science scores.

    a)   The confidence interval is: $(\bar{y}_1 - \bar{y}_2) \pm t^*_{df} \times SE(\bar{y}_1 - \bar{y}_2)$ where the standard error of the difference of the means is given as 1.22. Calculate the observed difference in the two means: $\bar{y}_{2000} - \bar{y}_{1996} = 147 - 150 = -3$. We can assume that the degrees of freedom ≥ 7536 which is the df of the smaller sample. The critical t value for $df \geq 7536$ is 1.96 (from technology or estimate from table). The 95% confidence interval is: $-3 \pm 1.96 * 1.22 = -3 \pm 2.39 = (-0.61, -5.39)$. We are 95% confident that the mean score has declined from 1996 to 2000. The decline is lower by 0.61 to 5.39 points. Zero is not contained in the interval; therefore, the difference is significant.

b) Both samples are very large which makes the standard errors of these samples very small. The result should be accurate despite the sample size.

61. The Internet.

    a) The differences that were observed between the group of students with Internet access and those without were too great to be attributed to natural sampling variation.

    b) If their conclusion that the difference is significant is incorrect, they would reject the null hypothesis when it is actually true. This is a Type I error.

    c) No. this evidence doesn't prove that using the Internet at home can improve a student's science scores. There could be many other contributing factors to the increase in science scores. Perhaps those who have Internet access have other advantages as well.

63. Pizza sales.

    a) The difference between the mean amount of this brand of frozen pizza sold between the two seasons is $31,234 - 22,475 = 8759$ pounds.

    b) The confidence interval is: $(\bar{y}_1 - \bar{y}_2) \pm t^*_{df} \times \text{SE}(\bar{y}_1 - \bar{y}_2)$ where the standard error of the difference

        of the means is: $\text{SE}(\bar{y}_W - \bar{y}_S) = \sqrt{\dfrac{s_1^2}{n_1} + \dfrac{s_2^2}{n_2}}$ Calculate the observed difference in the two means:

        $\bar{y}_W - \bar{y}_S = 8,759$. $\text{SE}(\bar{y}_W - \bar{y}_S) = \sqrt{\dfrac{(13500)^2}{38} + \dfrac{(8442)^2}{40}} = 2564.71$. From technology, the degrees

        of freedom are 61.5. The critical t value for df of 61.5 is 1.9996 (manually calculate or from technology). The 95% confidence interval is:

        $8759 \pm 1.9996 * 2564.71 = 8759 \pm 5128.39 = (3630.54, 13,887.39)$. Answer from technology may be slightly different due to roundoff in manual calculations. We are 95% confident that the interval 3,631 to 13,887 pounds contains the true difference in mean sales between winter and summer.

    c) Weather and sporting events may impact pizza sales.

65. Olympic heats. The hypotheses are:
$H_0 : \mu_2 - \mu_5 = 0;\ H_A : \mu_2 - \mu_5 \ne 0$.
Independent groups assumption: the two heats were independent. Randomization condition: runners were randomly assigned. Nearly Normal condition: boxplots show and outlier in the distribution of times in heat 2. Recommend to perform the test twice, once with the outlier and once without.

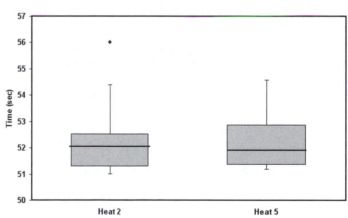

With the outlier, using technology, $t = 0.035$ with $df = 10.82$. This results in a P-value of 0.973 which is an extremely large number that gives support to the null hypothesis that states there is no difference in the means between the two heats. We fail to reject the null hypothesis.

Without the outlier in Heat 2, using technology, $t = -1.14$ with $df = 8.83$. This results in a P-value of 0.287 which is still > 5% and still fails to reject the null hypothesis.

Regardless of whether the outlier is included or removed, there is not evidence of a difference in the mean times between Heat 2 and Heat 5.

67. Tee tests. The hypotheses are: $H_0 : \mu_S - \mu_R = 0$; $H_A : \mu_S - \mu_R > 0$. Assuming the conditions are satisfied, it is appropriate to use t procedures. From technology, $t = 4.57$ with $df = 7.03$. The resulting P-value is 0.0013 which is $< 5\%$ and small enough to reject the null hypothesis. There is strong evidence that the mean ball velocity for Stinger tees is higher than the mean velocity for regular tees.

69. Marketing slogan.

    a)   The hypotheses are: $H_0 : \mu_M - \mu_R = 0$; $H_A : \mu_M - \mu_R > 0$. Independent groups assumption: the groups are not related with regards to memory score. Randomization condition: subjects were randomly assigned to groups. Nearly Normal condition: we don't have access to the actual data. We will assume that the distributions of the populations of memory test scores are Normal. From technology, $t = -0.70$ with $df = 45.88$. The resulting P-value 0.756 which is $> 5\%$ and large enough to support the null hypothesis. There is insufficient evidence to believe that the mean number of objects remembered by those who listen to Mozart is higher than the mean number of objects remembered by those who listen to rap music.

    b)   From technology, we are 90% confident that the mean number of objects remembered by those who listen to Mozart is between 5.28 and 0.26 lower than those who listened to no music.

71. Mutual funds returns.

    a)   Paired data assumption is met: the same 35 funds are looked at in both instances. Randomization condition: sample was taken randomly. Nearly Normal condition: the distributions are approximately symmetric with no outliers.

    b)   The hypotheses are: $H_0 : \mu_5 - \mu_3 = 0$; $H_A : \mu_5 - \mu_3 > 0$.

    c)   Using technology to perform the paired t-test, $t = 15.48$ with 34 df and P-value < 0.001. Therefore we can reject the null hypothesis and conclude that there is sufficient evidence to suggest that the average return is greater over 5 years compared to 3 years.

    d)   The 95% confidence interval for the mean difference is (4.224, 5.501).

73. Real estate, two towns.

    a) Independent groups assumption: the samples are not related. Randomization condition: sample was taken randomly. Nearly Normal condition: the distributions are approximately symmetric with no outliers. We can safely use the two-sample *t*-procedures.

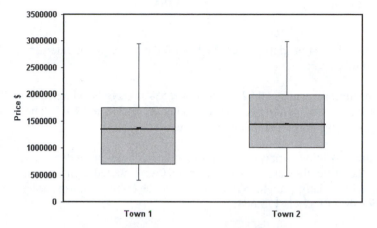

    b) The hypotheses are: $H_0 : \mu_1 - \mu_2 = 0; H_A : \mu_1 - \mu_2 \neq 0$.

    c) From technology, $t = -0.58$ with $df = 27.26$. The resulting P-value = 0.566 which is > 5% so we fail to reject the null hypothesis.

    d) We conclude that the mean price of homes in these two towns is not significantly different.

75. Designated hitter, part 3.

    a) Independent groups assumption: the samples are not related. Randomization condition: sample was taken randomly. Nearly Normal condition: the distributions are approximately symmetric with no outliers. We can safely use the two-sample *t*-procedures.

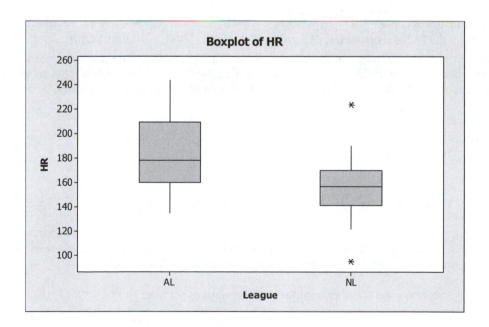

    b) The hypotheses are: $H_0 : \mu_{AL} - \mu_{NL} = 0; H_A : \mu_{AL} - \mu_{NL} > 0$.

c)   From technology, $t = 2.48$ with $df = 27$. The resulting P-value $= 0.010$.

d)   The P-value is < 0.05 so we reject the null hypothesis. There is sufficient evidence to conclude that the average number of home runs hit per game is higher in the American League compared to the National League.

77. Labor force.
a)   The paired t-test is appropriate. The labor force participation rate for two different years was paired by city.

b)   The resulting P-value is 0.0244 which is $< 5\%$ so there is evidence of a difference in the average labor force participation rate for women between 1968 and 1972. The evidence suggests an increase in the participation rate for women.

79. Online insurance. Adding variances requires that the variables are independent. The price quotes are for the same cars, therefore, they are paired. Drivers that were quoted high insurance premiums by the local company would also likely get a high rate quote from the online company. The standard deviation of the column of price differences is 175.663.

81. Online insurance, part 2.
a)   The histogram gives us better information on the differences in price. The histogram is compiled from data of *Local – Online*. Zero represents no difference and positive values represent a positive difference of *Local – Online*. The bigger the difference, the lower the online price.

b)   Insurance quotes are based on a number of factors of risk such as age, sex, previous accidents. Drivers are likely to get similar quotes both locally and online making the differences and the spread or standard deviation smaller also.

c)   The price quotes are paired (same driver for both types of companies). The random sample consisted of fewer than 10% of the agent's customers. The histogram of differences is approximately Normal.

83. Online insurance, part 3. The hypotheses test the one sample representing the difference between the Local and the Online companies: $H_0 : \mu(\text{Local - Online}) = 0$; $H_A : \mu(\text{Local - Online}) > 0$. No difference results in zero, the definition of the null hypothesis. Saving money would result in a positive difference. From technology, $t = 0.826$ with $df = 9$. The resulting P-value $= 0.215$ which is $> 5\%$ so we fail to reject the null hypothesis. These data do not provide evidence that online premiums are lower, on average.

85. Employee athletes.
a)   The hypotheses test the one sample representing the difference between before and after program: $H_0 : \mu_{Diff} = 0$; $H_A : \mu_{Diff} \neq 0$.

b)   The t-value of the difference: The observed t-value is: $t = \dfrac{(\overline{y}_{Diff})}{SE(\overline{y}_{Diff})} = \dfrac{22.7}{\dfrac{113.6}{\sqrt{145}}} = 2.406$ resulting in

a P-value of 0.017 which is $< 5\%$ and small enough to reject the null hypothesis. We conclude that the mean number of keystrokes per hour has changed (increased).

c)   The 95% confidence interval for mean keystrokes per hour is:

$22.7 \pm t_{0.025,144} * \dfrac{s}{\sqrt{n}} = 22.7 \pm 1.977 * \dfrac{113.6}{\sqrt{145}} = 22.7 \pm 18.65 = (4.05, 41.35)$.

87. Exercise equipment.

a) The Paired data assumption: the data are paired by type of exercise machine. Randomization condition: assume that the men and women participating are representative of all men and women in terms of the number of minutes of exercise required to burn 200 calories. 10% condition: the participant number is less than is less than 10% of all people. Nearly Normal condition: The histogram of the distribution of differences is at least roughly symmetric and unimodal. From technology, we are 95% confident that women take an average of 4.8 to 15.2 minutes longer than men to burn 200 calories at a light exertion rate.

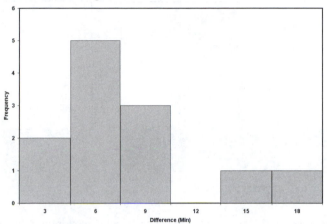

b) Nearly Normal condition: there is nothing unusual about the histogram of the differences and no reason to believe that they do not come from a Normal population. From technology, we are 95% confident that women exercising with light exertion take an average of 4.9 to 20.4 minutes longer to burn 200 calories than women exercising with hard exertion.

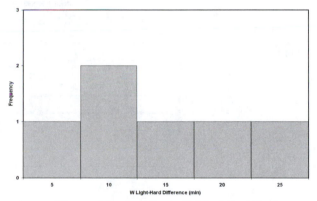

c) These data are averages. We would expect the individual times to be more variable. Therefore, our standard error would be larger, resulting in a larger margin of error.

89. Stopping distance.

a) Independent groups assumption: the wet and dry pavement stops were made under different conditions and were not paired in any way. Randomization condition: the stops can be assumed to be representative of these types of stops for this type of car, but not all cars (differences in types of brakes, etc.). 10% condition: 10 stops are less than 10% of all possible stops. Nearly Normal condition: histograms of the distributions of stopping distances are roughly symmetric and unimodal. The wet pavement is slightly left skewed.

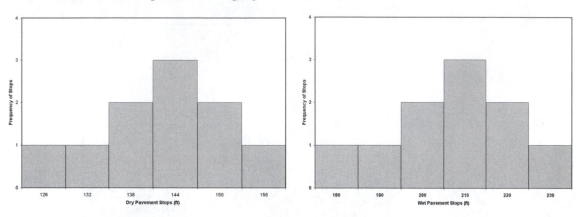

We are 95% confidence that the mean dry pavement stopping distance for this type of car is between 133.6 and 145.2 feet.

b) The assumptions are the same as a). A normal quantile plot for the wet pavement stops shows a fairly Normal distribution.

We are 95% confidence that the increase in stopping distances from 60 mph is between 51.3 and 74.7 feet.

91. Airlines.

a) The percent change in airfare does represent paired data because the locations and airfares are compared for two time periods.

b) The test is conducted as a paired *t*-test: $H_0 : \mu_D = 0$; $H_A : \mu_D \neq 0$. Mean difference (3Q2006 − 3Q2007) = 10.10; $t = 0.67$; P-value = 0.535; do not reject the null hypothesis. There is no evidence that the third quarter 2006 airfares are significantly different from the third quarter 2007 airfares. For the percentage differences, $t = 0.67$ and the P-value = 0.520. There is not enough evidence to support that the percent change between the quarters is different from zero.
Both tests provide the same statistical conclusion. However, in general, if there are extreme or outlying paired observations that skewed the mean difference, then the mean percent change might be a more appropriate test because it would eliminate the variability due to extreme differences.

Chapter 15 – Inference for Counts: Chi-Square Tests

SECTION 15.1

1.

a) If there is no "seasonal effect" on births we would expect an equal number of births for each season or $120/4 = 30$ births in winter, spring, summer and fall.

b) $$\chi^2 = \sum_{allcells} \frac{(Obs - Exp)^2}{Exp}$$

$$\chi^2 = \frac{(25-30)^2}{30} + \frac{(35-30)^2}{30} + \frac{(32-30)^2}{30} + \frac{(28-30)^2}{30}$$

$$\chi^2 = 0.8333 + 0.8333 + 0.1333 + 0.1333 = 1.9332$$

c) $Df = 4\text{-}1 = 3$.

SECTION 15.2

3.

a) The expected value (mean) of a χ^2 is its number of degrees of freedom. In this case it is 3.
b) The computed test statistic value of 1.933 is smaller than the mean.
c) It suggests that we cannot reject the null hypothesis that states that there is no seasonal effect on births (that births are distributed uniformly over the four seasons).
d) Using Table X, find the χ^2 value that corresponds to 3 df and a right-tail probability of 0.05. Therefore, $\chi^2 = 7.815$.
e) At $\alpha = 0.05$, we cannot reject the null hypothesis. Therefore, there is insufficient evidence to conclude that births are not uniformly distributed across seasons. Or alternatively, there is insufficient evidence to suggest a seasonal effect on births.

SECTION 15.3

5.

a) To compute the standardized residual for each season use $$\frac{(Obs - Exp)}{\sqrt{Exp}}.$$

For winter we have $\frac{(25-30)}{\sqrt{30}} = -0.913$; for spring we have $\frac{(35-30)}{\sqrt{30}} = 0.913$; for summer we have

$\frac{(32-30)}{\sqrt{30}} = 0.365$; and for fall we have $\frac{(28-30)}{\sqrt{30}} = -0.365$.

b) None of the standardized residuals are particularly large when viewed as $z-$ scores.
c) We would have anticipated that none of the standardized residuals would be too large because we did not reject the null hypothesis.

SECTION 15.4

7.

a) The null hypothesis is that the age distribution of customers is the same at both branches of the local bank.

b)   The appropriate test is the Chi-Square test of homogeneity.

c)   The expected frequencies are shown in the table below:

Age	In town branch	Mall branch	Total
Younger than 30	25	25	50
30 to 55	45	45	90
56 or Older	30	30	60
Total	100	100	200

d) $\chi^2 = \displaystyle\sum_{allcells} \frac{(Obs - Exp)^2}{Exp}$

$$\chi^2 = \frac{(20-25)^2}{25} + \frac{(40-45)^2}{45} + \frac{(40-30)^2}{30} + \frac{(30-25)^2}{25} + \frac{(50-45)^2}{45} + \frac{(20-30)^2}{30} = 9.778$$

e)   The degrees of freedom are $(R-1) \times (C-1) = (3-1) \times (2-1) = 2$.

f)   Using Table X, find the χ^2 value that corresponds to 2 *df* and a right-tail probability of 0.05. Therefore, $\chi^2 = 5.991$.

g)   At $\alpha = 0.05$, we can reject the null hypothesis. Therefore, there is sufficient evidence to conclude that the age distribution of customers is different at the two branches of the local bank.

SECTION 15.5

9.

a)   The proportion of men who said they use the Internet occasionally is $\hat{p}_1 = 2393/3037 = 0.7879$ and the proportion of women who said they use the Internet occasionally is $\hat{p}_2 = 2378/3166 = 0.7511$.

b)   The difference is $\hat{p}_1 - \hat{p}_2 = 0.7879 - 0.7511 = 0.0368.$.

c)   The standard error of the difference is

$$SE(\hat{p}_1 - \hat{p}_2) = \sqrt{\frac{\hat{p}_1\hat{q}_1}{n_1} + \frac{\hat{p}_2\hat{q}_2}{n_2}} = \sqrt{\frac{(.7879)(.2121)}{3037} + \frac{(.7511)(.2489)}{3166}} = 0.0107$$

d)   The 95% confidence interval for the difference in proportions is

$(\hat{p}_1 - \hat{p}_2) \pm z^* SE(\hat{p}_1 - \hat{p}_2)$

$0.0368 \pm (1.96)(0.0107)$

$(0.0158, 0.0578)$ or 1.58% to 5.78%.

SECTION 15.6

11.

a)   The expected values under the null hypothesis are shown in the table below:

Did You Use the Internet Yesterday?	White	Black	Hispanic/Other
Yes	2506.38	338.90	445.73
No	895.62	121.10	159.27

b)

$$\chi^2 = \sum_{allcells} \frac{(Obs - Exp)^2}{Exp}$$

$$\chi^2 = \frac{(2546 - 2506.38)^2}{2506.38} + \frac{(314 - 338.90)^2}{338.90} + \frac{(431 - 445.73)^2}{445.73} +$$

$$\frac{(856 - 895.62)^2}{895.62} + \frac{(146 - 121.10)^2}{121.10} + \frac{(174 - 159.27)^2}{159.27} = 11.176.$$

   c)   The degrees of freedom are (R-1) × (C-1) = (2-1) × (3-1) = 2.
   d)   Using Table X, find the χ^2 value that corresponds to 1 *df* and a right-tail probability of 0.05. Therefore, $\chi^2 = 5.991$.
   e)   Reject the null hypothesis and conclude that Internet usage is not independent of racial background.

CHAPTER EXERCISES

13. Concepts.
   a)   Chi-square test of independence. One sample, two variables. We want to see if the variable *Account type* is independent of the variable *Trade type*.

   b)   Some other statistical test; the variable *Account size* is quantitative, not counts.

   c)   Chi-square test of homogeneity; we have two samples (residential and nonresidential students) and one variable, *Courses*. We want to see if the distribution of *Courses* is the same for both groups.

15. Dice.
   a)   If the die were fair, you'd expect each face to show 10 times.

   b)   Use a chi-square test for goodness-of-fit. We are comparing the distribution of a single variable (outcome of a die roll) to an expected distribution.

   c) H_0: The die is fair (all faces have the same probability of coming up).
        H_A: The die is not fair (some faces are more or less likely to come up than others).

   d)   **Counted data condition**: We are counting the number of times each face comes up.
        Randomization condition: Die rolls are random and independent of each other.
        Expected cell frequency condition: We expect each face to come up 10 times, and 10 > 5.

   e)   Under these conditions, the sampling distribution of the test statistic is χ^2 on 6-1 = 5 degrees of freedom. We will use a chi-square goodness-of-fit test.

   f)

Face	Observed	Expected	Residual = $(Obs - Exp)$	$(Obs - Exp)^2$	Component = $\dfrac{(Obs - Exp)^2}{Exp}$
1	11	10	1	1	0.1
2	7	10	− 3	9	0.9
3	9	10	− 1	1	0.1
4	15	10	5	25	2.5
5	12	10	2	4	0.4
6	6	10	− 4	16	1.6

$$\sum = 5.6$$

g) Since the P-value = 0.3471 is high, we fail to reject the null hypothesis. There is no evidence that the die is unfair.

17. Quality control, part 2.

a) The weights of the nuts are quantitative. Chi-square goodness-of-fit requires counts.

b) In order to use a chi-square test, you could count the number of each type of nut. However, it's not clear whether the company's claim was a percentage by number or a percentage by weight.

19. Maryland lottery.

a) Use a chi-square test for goodness-of-fit. We are comparing the distribution of a single variable (Count) to an expected distribution.

b) **Counted data condition**: Counts are actual lottery counts.
Randomization condition: The lottery mechanism uses randomization and guarantees independence.
Expected cell frequency condition: The expected counts are all greater than 5.

c) H_0: The likelihood of drawing each numeral is equal.
H_A: The likelihood of drawing each numeral is *not* equal.

d)

Group	Observed	Expected	Residual = (Obs - Exp)	(Obs - Exp)^2	(Obs - Exp)^2/Exp
0	62	65.4	-3.4	11.56	0.18
1	55	65.4	-10.4	108.16	1.65
2	66	65.4	0.6	0.36	0.01
3	64	65.4	-1.4	1.96	0.03
4	75	65.4	9.6	92.16	1.41
5	57	65.4	-8.4	70.56	1.08
6	71	65.4	5.6	31.36	0.48
7	74	65.4	8.6	73.96	1.13
8	69	65.4	3.6	12.96	0.20
9	61	65.4	-4.4	19.36	0.30

$$\sum \approx 6.46$$

$\chi^2 = 6.46$. The resulting P-value is 0.693 and too large to reject the null hypothesis H_0.

e) The P-value says that if the drawings were fair, an observed chi-square value of 6.46 of higher would occur about 69% of the time. This is not unusual at all, so we don't reject the null hypothesis that the values are uniformly distributed. The variation that is observed is typical of what is expected if the digits were equally likely.

21. Titanic.

a) $P(\text{crew}) = \dfrac{885}{2201} \approx 0.402$

b) $P(\text{third and alive}) = \dfrac{178}{2201} \approx 0.081$

c) $P(\text{alive} \mid \text{first}) = \dfrac{P(\text{alive and first})}{P(\text{first})} = \dfrac{202/2201}{325/2201} = \dfrac{202}{325} \approx 0.622$

d) The overall chance of survival is $\dfrac{710}{2201} \approx 0.323$, so we would expect about 32.3% of the crew, or about 285.48 members of the crew, to survive.

e) Chi-square test of independence with hypotheses:

H_0: Survival was independent of status on the ship.

H_A: Survival depended on status on the ship.

f) The table has 2 rows and 4 columns, so there are $(2-1) \times (4-1) = 3$ degrees of freedom.

g) With $\chi^2 \approx 187.8$, on 3 degrees of freedom, the P-value is essentially 0, so we reject the null hypothesis. There is strong evidence survival depended on status. First-class passengers were more likely to survive than any other class or crew.

23. Birth order and college choice.

a) This is a chi-square test for independence. There is one sample of students, and two variables, *birth order* and *college*. We want to know if *college* choice is independent of *birth order*.

b) H_0: College enrollment is independent of birth order.

H_A: There is an association between college enrollment and birth order.

c) **Counted data condition:** The data are counts.
Randomization condition: This is not a random sample of students, but there is no reason to think that this group of students isn't representative, at least of students in a statistics class.
Expected cell frequency condition: The expected counts are low for both the Social Science and Professional Colleges for both third and fourth or higher birth order. We'll keep and eye on these when we calculate the standardized residuals.

d) There are 4 rows and 4 columns, for (4-1)(4-1) = 9 degrees of freedom.

e) With a P-value this low, we reject the null hypothesis. There is some evidence of an association between birth order and college enrollment.

f) Unfortunately, 3 of the 4 largest standardized residuals are in cells with expected counts less than 5. We should be very wary of drawing conclusions from this test.

25. Cranberry juice.

a) We are concerned with the proportion of urinary tract infections among three different groups. We will use a chi-square test for homogeneity.

b) **Counted data condition**: The data are counts.
Randomization condition: Although not specifically stated, we will assume that the women were randomly assigned to treatments.
Expected cell frequency condition: The expected frequencies are all greater than 5.

c) H_0: The proportion of urinary tract infections is the same for each group.
H_A: The proportion of urinary tract infections is different among the groups.

d) The table has 2 rows and 3 columns, so there are (2-1) $\times$ (3-1) = 2 degrees of freedom.

$$\chi^2 = \sum_{all\ cells} \frac{(Obs - Exp)^2}{Exp} \approx 7.776$$

P-value ≈ 0.020.

e)   Since the P-value is low, we reject the null hypothesis. There is strong evidence of a difference in the proportion of urinary tract infections for cranberry juice drinkers, lactobacillus drinkers, and women that drink neither of the two beverages.

f)   A table of the standardized residuals is below, calculated by using $c = \dfrac{Obs - Exp}{\sqrt{Exp}}$

	Cranberry	Lactobacillus	Control
Infection	−1.87276	1.191759	0.681005
No infection	1.245505	−0.79259	−0.45291

There is evidence that women who drink cranberry juice are less likely to develop urinary tract infections, and women who drank lactobacillus are more likely to develop urinary tract infections.

27. Market segmentation.

a)   This is a chi-squared test of independence. There is one sample of customers and two variables, *Age* and *Shopping Frequency*. We want to know if *Shopping Frequency* is independent of *Age*.

b) H_0: Age is independent of frequency of shopping at this department store.
H_A: Age is not independent of frequency of shopping at this department store.

c)   **Counted data condition**: The counts are recorded from a survey.
Randomization condition: Assume the survey was conducted randomly (not specifically stated).
Expected cell frequency condition: The expected frequencies are all greater than 5.

d)   Since the P-value is low, we reject the null hypothesis. There is evidence of an association between age and frequency of shopping at this department store.

e)   Given the negative residuals for the low frequency categories among the older women and the positive residuals for the higher frequency categories among the older women, we conclude that the older women in this survey shop more frequently at this department store than expected.

29. Shopping.

a)   The proportion of all adults in the sample that bought books online is 84/430 = 19.535%.
The observed counts: The expected counts:

Sex	Books Online	Books Not Online	Total
Men	47	175	222
Women	37	171	208
Total	84	346	430

Sex	Books Online	Books Not Online	Total
Men	43.37	178.63	222
Women	40.63	167.37	208
Total	84	346	430

$\chi^2 = 0.782$. The resulting *P*-value is 0.377 and too large to reject the null hypothesis H_0. There is not enough evidence to conclude that either men or women are more likely to make online purchases of books.

b)   Type II error (the error of failing to reject a null hypothesis when it is in fact false).

c) $\hat{p}_1 = 47/222 = 0.2117$ (Men) and $\hat{p}_2 = 37/208 = 0.1779$ (Women).
The 95% confidence interval is:

$$(\hat{p}_1 - \hat{p}_2) \pm z^* SE(\hat{p}_1 - \hat{p}_2) = (0.2117 - 0.1779) \pm 1.96 \times \sqrt{\dfrac{(0.2117)(0.7883)}{222} + \dfrac{(0.1779)(0.8221)}{208}}$$

$$= 0.0338 \pm 1.96 \times 0.0381 = 0.0338 \pm 0.0747 = (-4.09\%, 10.85\%)$$

31. Fast food.

  a) The proportion of all adults in the sample that agree is 443/800 = 55.38%.
    The observed counts: The expected counts:

Age	Agree	Not	Total
< or = 35	197	214	411
> 35	246	143	389
Total	443	357	800

Type	Agree	Not	Total
< or = 35	227.59	183.41	411
> 35	215.41	173.59	389
Total	443	357	800

  $\chi^2 = 18.95$. The resulting P-value is essentially zero and small enough to reject the null hypothesis H_0. There is evidence to conclude that the proportions are not equal.

  b) Type II error (the error of failing to reject a null hypothesis when it is in fact false).

  c) $\hat{p}_1 = 246/389 = 0.6323$ (Agree > 35) and $\hat{p}_2 = 197/411 = 0.4793$ (Agree < = 35).
    The 90% confidence interval is:

  $$(\hat{p}_1 - \hat{p}_2) \pm z^* SE(\hat{p}_1 - \hat{p}_2) = (0.6323 - 0.4793) \pm 1.645 \times \sqrt{\frac{(0.6323)(0.3677)}{389} + \frac{(0.4793)(0.5207)}{411}}$$

  $$= 0.1530 \pm 1.645 \times 0.0347 = 0.1530 \pm 0.0571 = (9.59\%, 21.01\%)$$

33. Foreclosure rates.

  a) The proportion of all houses in sample in foreclosure is 14/2558 = 0.55%.
    The observed counts: The expected counts:

State	Foreclosures	Not	Total
Nevada	8	1090	1098
Colorado	6	1454	1460
Total	14	2544	2558

State	Foreclosures	Not	Total
Nevada	6.01	1,091.99	1098
Colorado	7.99	1,452.01	1460
Total	14	2544	2558

  $\chi^2 = 1.162$. The resulting P-value is 0.281 and too large to reject the null hypothesis H_0. There does not seem to be a difference in the foreclosure proportions for the two states.

  b) $\hat{p}_1 = 8/1098 = 0.0073$ (Nevada) and $\hat{p}_2 = 6/1460 = 0.0041$ (Colorado).
    The 90% confidence interval is:

  $$(\hat{p}_1 - \hat{p}_2) \pm z^* SE(\hat{p}_1 - \hat{p}_2) = (0.0073 - 0.0041) \pm 1.645 \times \sqrt{\frac{(0.0073)(0.9927)}{1098} + \frac{(0.0041)(0.9959)}{1460}}$$

  $$= 0.0032 \pm 1.645 \times 0.0031 = 0.0032 \pm 0.0051 = (-0.19\%, 0.83\%)$$

35. Market segmentation, part 2.

  H_0: Marital status is independent of frequency of shopping.
  H_A: Marital status is not independent of frequency of shopping.

  Counted data condition: The counts are recorded from a survey.
  Randomization condition: Assume the survey was conducted randomly (not specifically stated).
  Expected cell frequency condition: The expected counts are all greater than 5.
  $\chi^2 = 23.858$. The resulting P-value is 0.001 and is small enough to reject the null hypothesis H_0.
  There is strong evidence of an association between marital status and frequency of shopping at this department store. Based on the residuals, married customers shopped at this store more frequency than expected, and more single women shopped never/hardly ever than expected.

37. Accounting.

  a) Chi-square test for homogeneity. We have two samples (2000 and 2006) and one variable, *Answer to one question*. We want to see if the distribution of *answers to the question* is the same for both years.

b) **Counted data condition**: The counts are recorded from a survey.
Randomization condition: The executives were surveyed randomly.
Expected cell frequency condition: The expected counts are all greater than 5.

c) H_0: The distribution of attitudes about critical factors affecting ethical and legal accounting practices was the same in 2000 and 2006.
H_A: The distribution of attitudes about critical factors affecting ethical and legal accounting practices was not the same in 2000 and 2006.

d) $\chi^2 = 4.03$. The resulting P-value is 0.402.

e) Since the P-value is high, we cannot reject the null hypothesis H_0. There is no evidence of an association between the attitudes about factors affecting ethical and legal accounting practices and the span between 2000 and 2006.

39. Market segmentation, part 3.

a) Chi-square test of independence. There is one sample of customers and two variables, *Quality* and *Shopping Frequency*. We want to know if *Shopping Frequency* is independent of *Quality*.

b) **Counted data condition**: The counts are recorded from a survey.
Randomization condition: Assume the survey was conducted randomly (not specifically stated).
Expected cell frequency condition: The expected counts are all greater than 5.

c) H_0: The emphasis on quality is independent of frequency of shopping.
H_A: The emphasis on quality is not independent of frequency of shopping.

d) $\chi^2 = 30.007$ with a *df* of 6. The resulting P-value is essentially zero (P < 0.001).

e) Since the P-value is low, we can reject the null hypothesis H_0. There is evidence of an association between the emphasis on quality and frequency of shopping at this department store.

41. Entrepreneurial executives, again.

a) Expected counts:

Perceived Value	Men	Women
Excellent	6.67	5.33
Good	12.78	10.22
Average	12.22	9.78
Below Average	8.33	6.67

A chi-test is appropriate because all of the counts are greater than 5. In addition, these executives are representative of all executives who completed the program.

b) The *df* decreased from 4 to 3. (r-1)(c-1) = (4-1)(2-1) = 3

c) $\chi^2 = 9.306$. The resulting P-value = 0.0255. Since the P-value is low, we reject the null hypothesis. There is evidence that the distributions of responses about the value of the program for men and women executives are different.

43. Racial steering.

H_0: There is no association between race and section of the complex in which people live.

H_A: There is an association between race and section of the complex in which people live.

Counted data condition: The data are counts.
Randomization condition: Assume that the recently rented apartments are representative of all apartments in the complex.

	White (Obs / Exp)	Black (Obs / Exp)
Section A	87 / 76.179	8 / 18.821
Section B	83 / 93.821	34 / 23.179

Expected cell frequency condition: The expected counts are all greater than 5.

Under these conditions, the sampling distribution of the test statistic is χ^2 on 1 degree of freedom. We will use a chi-square test for independence.

$$\chi^2 = \sum_{all\ cells} \frac{(Obs - Exp)^2}{Exp} \approx \frac{(87 - 76.179)^2}{76.179} + \frac{(8 - 18.821)^2}{18.821} + \frac{(83 - 93.821)^2}{93.821} + \frac{(34 - 23.179)^2}{23.179}$$

$$\approx 1.5371 + 6.2215 + 1.2481 + 5.0517$$

$$\approx 14.058$$

With $\chi^2 \approx 14.058$, on 1 degree of freedom, the P-value ≈ 0.0002.

Since the P-value ≈ 0.0002 is low, we reject the null hypothesis. There is strong evidence of an association between race and the section of the apartment complex in which people live. An examination of the components shows us that whites are much more likely to rent in Section A (component = 6.2215), and blacks are much more likely to rent in Section B (component = 5.0517).

45. Racial steering, revisited.

Section A: $\hat{p}_A = 8/95 = 0.0842$ Section B: $\hat{p}_B = 34/117 = 0.2906$

$\hat{p}_B - \hat{p}_A = 0.2064$

The 95% confidence interval is:

$$(\hat{p}_B - \hat{p}_A) \pm z^* SE(\hat{p}_B - \hat{p}_A) = 0.2064 \pm 1.96 \times \sqrt{\frac{(0.0842)(0.9158)}{95} + \frac{(0.2906)(0.7094)}{117}}$$

$$= 0.2064 \pm 1.96 \times 0.0507 = 0.2064 \pm 0.0994 = (10.70\%, 30.58\%)$$

47. Industry sector and outsourcing.

a) Chi-square test of independence. There is one sample of companies and two variables, *Industry Sector* and *Level of Outsourcing*. We want to know if *Industry Sector* is independent of *the Level of Outsourcing*.

b) **Counted data condition**: The counts are recorded from a survey.
Randomization condition: Assume the sample was taken randomly (not specifically stated).
Expected cell frequency condition: The expected counts are all greater than 5.

c) H_0: Outsourcing is independent of industry sector.
H_A: There is an association between outsourcing and industry sector.

d) $\chi^2 = 2815.97$ with a *df* of 9. The resulting *P*-value is essentially zero.

e) Since the P-value is so low, we reject the null hypothesis. There is strong evidence of an association between outsourcing and industry sector.

49. Management styles.

a)   Chi-square test for homogeneity. (Could be independence if the categories are exhaustive)

b)   **Counted data condition**: The counts are recorded from a survey.
Randomization condition: Assume the sample was taken randomly (not specifically stated).
Expected cell frequency condition: The expected counts are all greater than 5.

c) H_0: The distribution of employee job satisfaction level attained is the same for different management styles.
H_A: The distribution of employee job satisfaction level attained is different for different management styles.

d) $\chi^2 = 178.453$ with a df of 12. The resulting P-value is essentially zero.

e)   Since the P-value is so low, we reject the null hypothesis. There is strong evidence of an association between employee job satisfaction level and different management styles. Generally, exploitive authoritarian styles are more likely to have lower employee job satisfaction than consultative or participative styles.

51. Businesses and blogs. Assumptions and conditions for test of independence satisfied.
H_0: Reading online journals or blogs is independent of generation.
H_A: There is an association between reading online journals or blogs and generation.

$\chi^2 = 48.408$ for a df of 8. The resulting P-value is essentially zero and < 0.001.

We reject the null hypothesis and conclude that reading online journals or blogs is not independent of generational age.

53. Information systems. Use a Chi-square test of homogeneity (unless these two types of firms are considered the only two types in which case it's a test of independence).
Assumptions:
Counted data condition: Data provided is in counts.
Randomization condition: Assume that the sample was random although not stated.
Expected cell frequency condition: The expected counts are all greater than 5.

H_0: Systems used have same distribution for both types of industry.
H_A: Distributions of type of system differs in the two industries.

$\chi^2 = 157.256$ with a df of 3. The resulting P-value is essentially zero.

Since the P-value is low, we can reject the null hypothesis and conclude that the type of ERP system used differs across industry type. Those in manufacturing appear to use more of the inventory management and ROI systems.

55. Economic growth. We will use a chi-square test of independence to see if the percent change in GDP is independent of region.
Assumptions:
Counted data condition: Data provided is in counts.
Randomization condition: Assume that this time period is representative.
Expected cell frequency condition: The expected counts are less than 5 in three cells, but very close. We will proceed cautiously.

H_0: Economic growth is independent of region of the United States.
H_A: Economic growth is not independent of region of the United States.

$\chi^2 = 19.072$ with a df of 3. The resulting P-value < 0.001.

The P-value is so low, we reject the null hypothesis and conclude that economic growth is not independent of region. The result seems clear enough even though the expected counts of three cells were slightly below 5.

Chapter 16 – Inference for Regression

SECTION 16.1

1.

The amount spent eating out per week by people in this target community is about $17.28 plus $0.35 for each $1000 of their household income.

SECTION 16.2

3.

The scatterplot (below) shows a negative relationship between age of the website's customers and the number of movies rented online.

a)   Linearity: The linearity assumption is satisfied as the scatterplot appears linear.

b)   Independence: If we can assume that the customers are selected at random from the population of all customers to the website, then we can assume that the errors in the true underlying regression model are independent of each other.

c)   Equal spread: While it is best to check the equal spread condition using a residual plot, we can also examine the scatterplot for a consistent spread. In this case there is no indication of inconsistent spread and we can assume this condition is satisfied.

d)   Normality: We need residuals to check this assumption; consequently we can't tell if it is satisfied based on what we know.

SECTION 16.3

5.

a)   In order to predict *Rentals* for all six values of *Age,* use $\hat{y} = 18.9 - 0.260x$. The predicted values for each value of x are shown in the table below.

Age (x)	Rentals ($\hat{y}$)
35	9.8
40	8.5
50	5.9
65	2.0
40	8.5
30	11.1

b) Use $e_i = y_i - \hat{y}_i$ to find residuals as shown below.

Age (x)	$e_i = y_i - \hat{y}_i$
35	9 – 9.8 = -0.8
40	8 - 8.5 = -0.5
50	4 - 5.9 = -1.9
65	3 - 2.0 = 1.0
40	10 - 8.5 = 1.5
30	12 - 11.1 = 0.9

c) The residual standard deviation is

$$s_e = \sqrt{\frac{\sum(y-\hat{y})^2}{n-2}}$$

$$s_e = \sqrt{\frac{(-0.8)^2 + (-0.5)^2 + (-1.9)^2 + (1.0)^2 + (1.5)^2 + (0.9)^2}{6-2}} = 1.46$$

7.

In order to find the standard error of the regression slope, we need the spread around the line $s_e = 1.46$, spread of x values $s_x = 12.5167$, and sample size $n = 6$.

$$SE(b_1) = \frac{s_e}{s_x\sqrt{n-1}} = \frac{1.46}{12.5167\sqrt{5}} = 0.052$$

SECTION 16.4

9.

a) The t–statistic is

$$t = \frac{b_1 - \beta_1}{SE(b_1)} = \frac{0.265 - 0}{0.0454} = 5.837$$

b) The degrees of freedom are $n - 2 = 5 - 2 = 3$.

c) The P-value associated with $t = 5.837$ is approximately 0.01.

d) Since the P-value is less than $\alpha = 0.05$, we can reject the standard null hypothesis for the slope. We can conclude that there is a linear relationship between *Concession Sales per person* at a theatre and *Minutes before the movie begins*.

SECTION 16.5

11.

a) The null and alternative hypotheses are H_0: $\rho = 0$ vs. H_A: $\rho \neq 0$.

b) The test statistic is

$$t = r\sqrt{\frac{n-2}{1-r^2}} = 0.20\sqrt{\frac{98-2}{1-(0.20)^2}} = 2.00$$

c)   The critical t value, with 96 df, is approximately 1.984. Since the calculated t statistic is larger than the critical value, we can reject the null hypothesis and conclude that the true correlation is not zero.

SECTION 16.6

13.

a)   Based on the regression equation, the predicted sales on a day with 12 employees is 19.068 ($1000's) or $19, 068.

b)   To find the confidence interval, first find the standard error

$$SE(\hat{\mu}_v) = \sqrt{SE^2(b_1) \times (x_v - \bar{x})^2 + \frac{s_e^2}{n}}$$

$$SE(\hat{\mu}_v) = \sqrt{(0.0873)^2 \times (12 - 10.4)^2 + \frac{(1.477)^2}{10}} = 0.488$$

The 95% confidence interval is

$$\hat{y} \pm t^*_{n-2} \times SE = 19.068 \pm 2.306 \times 0.488$$

(17.94, 20.19) $1000's or $17,940 to $20,190.

c)   To find the prediction interval, first find the standard error

$$SE(\hat{y}_v) = \sqrt{SE^2(b_1) \times (x_v - \bar{x})^2 + \frac{s_e^2}{n} + s_e^2}$$

$$SE(\hat{y}_v) = \sqrt{(0.0873)^2 \times (12 - 10.4)^2 + \frac{(1.477)^2}{10} + (1.477)^2} = 1.553$$

The 95% prediction interval is

$$\hat{y} \pm t^*_{n-2} \times SE = 19.068 \pm 2.306 \times 1.553$$

(15.48, 22.66) $1000's or $15,480 to $22,660.

15.

a)   The restaurant owner should interpret this prediction interval as follows: 95% of potential customers in the community from households that have annual incomes of $80,000 can be expected to spend between $35.60 and $55.60 eating out each week. Therefore she should consider pricing meals in this range to attract more customers in this demographic.

b)   We can be 95% confident that the mean amount spent weekly by individuals whose household incomes are $80,000 is between $40.60 and $50.60. The mean is a summary for this demographic.

c)   Because we can estimate means more accurately than we can estimate individual values, confidence intervals for the mean are narrower than the corresponding prediction intervals for individual values.

SECTION 16.7

17.

a)   Based on the regression equation, the predicted value of songs when $Age = 16$ is 6.58.

b) To find the confidence interval for the mean, first find the standard error

$$SE(\hat{\mu}_v) = \sqrt{SE^2(b_1) \times (x_v - \bar{x})^2 + \frac{s_e^2}{n}}$$

$$SE(\hat{\mu}_v) = \sqrt{(0.182)^2 \times (16 - 16.2)^2 + \frac{(1.3)^2}{5}} = 0.583$$

The 95% confidence interval for the mean is

$$\hat{y} \pm t^*_{n-2} \times SE = 6.58 \pm 3.182 \times 0.583$$
(4.725, 8.435)

c) To find the prediction interval, first find the standard error

$$SE(\hat{y}_v) = \sqrt{SE^2(b_1) \times (x_v - \bar{x})^2 + \frac{s_e^2}{n} + s_e^2}$$

$$SE(\hat{y}_v) = \sqrt{(0.182)^2 \times (16 - 16.2)^2 + \frac{(1.3)^2}{5} + (1.3)^2} = 1.425$$

The 95% prediction interval is

$$\hat{y} \pm t^*_{n-2} \times SE = 6.58 \pm 3.182 \times 1.425$$
(2.045, 11.114)

d) The prediction interval is wider than the confidence interval. It reflects the greater amount of uncertainty present when predicting an individual value compared to predicting a mean.

CHAPTER EXERCISES

19. Online shopping.
  a) Scatterplot:

  b) The data plotted are at least moderately linear with a negative association. The data points are close together with some scatter towards the high end of age values. There is however evidence of nonlinearity at the high end of age values and slightly at the low end. A linear regression may not be completely appropriate for this data set.

c)   The regression line is: y = -11.503x + 750.02

d)   The residuals are shown below. The nearly normal condition is satisfied with the histogram. The residual scatterplot doesn't show any unusual patterns with the exception of some possible curvature. Any evidence of curvature may indicate we should proceed with caution using a linear model.

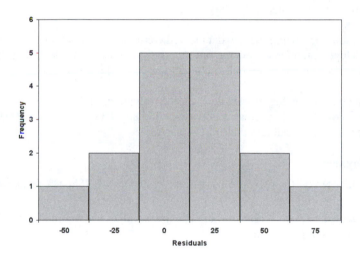

21. **Movie budgets.**

a)   The regression equation is: $\widehat{Budget} = -31.39 + 0.71\,Run\,Time$. The model suggests that for every one minute increase in time, the budget of a movie increases by 0.71 million dollars or $710,000.

b)   A negative intercept makes no physical sense in this case (representing the budget for a film that is zero minutes long). However, the P value of 0.0693 for the intercept indicates that we can't discern a difference between our estimated value and zero. It makes sense to say that a film of zero minutes in length costs $0 to make.

c)   The value of $s = 32.95$ is the residual standard deviation indicating that the amount by which movie budgets differ from predictions made by this model vary with a standard deviation of about $33 million.

d) The standard error of the slope is shown on the output as 0.15 $m/min.

e) The standard error of 0.15 $m/min can be interpreted as: for other models on a different sample of movies, we'd expect the slopes of the regression line to vary, with a standard deviation of about $150,000 per minute.

23. Movie budgets, part 2.
    a) The scatterplot looks reasonably straight. The residuals look random and roughly normal. The residuals don't display change in variability although there may be some increasing variability for higher age values (> 40).

    b) The 95% confidence interval for the slope (calculated from the raw data set) is between 0.41 and 1.01 million dollars per minute. The interpretation is that we are 95% confident that the cost of making movies increases at a rate of 0.41 to 1.01 million dollars per minute.

25. Water hardness.
    a) The hypotheses to test the association between the two variables:
        H_0: There is no linear relationship between calcium concentration in water and mortality rates for males. $(\beta_1 = 0)$.
        H_A: There is a linear relationship between calcium concentration in water and mortality rates for males. $(\beta_1 \neq 0)$.

    b) The t value = -6.65 indicating a P value of < 0.0001; reject the null hypothesis. There is strong evidence of a linear relationship between calcium concentration and mortality. Towns with higher calcium concentrations tend to have lower mortality rates.

    c) From technology using the raw data set, (the critical t value = 2.001) the 95% confidence interval for the slope (calculated from the raw data set) is between -4.196 and -2.256 deaths per 100,000 for each additional part per million of calcium in drinking water; (-4.196, -2.256).

    d) From technology using the raw data set, the interpretation is that we are 95% confident that the average mortality rate decreases between 2.256 and 4.196 deaths per 100,000 for each additional part per million of calcium in drinking water.

27. Youth unemployment.
    a) From technology, the regression equation: $\widehat{male\ rate} = 0.75564 * (female\ rate) + 2.376$.

    b) The scatterplot shows a high outlier. The lower values show a moderate positive linear association with a fair amount of scatter.

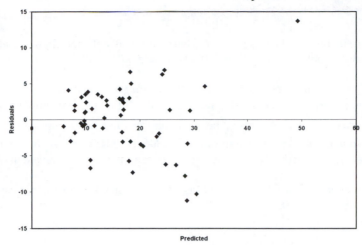

The residual plot does not show any unusual variation. In addition, the outlier does not seem to fit the other data points. The analysis should be run with and without the outlier to determine its effect on the model.

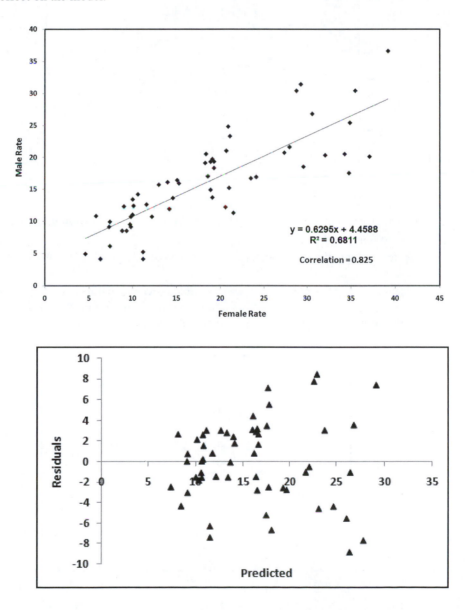

Without the outlier, the regression and correlation only change slightly and we determine that the outlier does not have a significant effect on the model

c) The hypotheses to test the association between the two variables:

H_0: There is no linear relationship between male and female unemployment rates. $(\beta_1 = 0)$

H_A: There is a linear relationship between male and female unemployment rates. $(\beta_1 \neq 0)$

$t = 13.46$ with $df = 55$, and the resulting P value < 0.001; reject the null hypothesis. There is strong evidence of a positive linear relationship between the male and female unemployment rates.

d) The percentage of variability in the Male Rate accounted for by the regression model is represented by $R^2 = 0.7671$ or 76.7%.

29. Used cars.

a) The scatterplot:

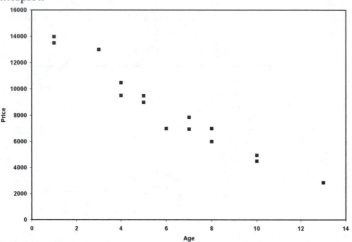

b) The scatterplot shows a fairly strong linear negative association without outliers, therefore, a linear model would be appropriate.

c) From technology, the regression equation: $\overline{Advertised\ Price} = 14,286 - 959 * (Age)$.

d) The scatterplot of the residuals shows a possible curvature and increased spread for higher values. The histogram of the residuals shows a fairly normal distribution. We will proceed with caution understanding that inference may not be entirely valid here.

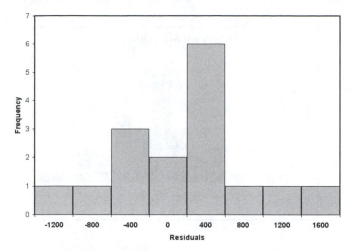

31. Used cars, part 2. 95% confidence interval:

$$b_1 \pm t^*_{n-2} \times SE(b_1) = -959 \pm 2.160 * 64.58 = -959 \pm 139.49 = (-\$1,098.5, -\$819.5)$$

Based on these data we are 95% confident that a used car's *Price* decreases between $819.50 and $1098.50 per year.

33. Fuel economy.

a)

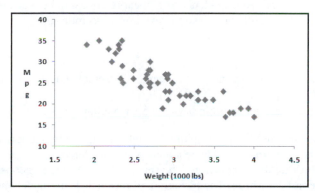

From technology, $\widehat{Mpg} = 48.739 - 8.2136 * Weight$

b) Conditions:
Linearity condition: The scatterplot shown below is fairly straight.
Independence assumption: The residuals plot is not curved.
Equal spread condition: The residuals show some increase in values as the predicted values increase. This might cause issues with linear regression model.
Nearly Normal condition: The histogram of the residuals is fairly symmetric with one possible outlier that is not extreme but could affect the linear model.

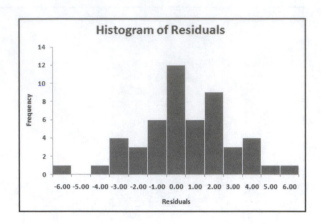

c) The hypotheses to test the association between the two variables:
    H_0: There is no linear relationship between the weight of a car and its fuel efficiency. $(\beta_1 = 0)$.

    H_A: There is a linear relationship between the weight of a car and its fuel efficiency. $(\beta_1 \neq 0)$.

d) Using technology, $t = -12.19$ with $df = 48$ resulting in a P-value < 0.0001. We reject the null hypothesis. There is strong evidence of a linear relationship between the weight of a car and its mileage. Cars that weigh more tend to have lower gas mileage.

35. SAT scores.

a) H_0: There is no linear relationship between SAT Verbal and Math scores. $\left(\beta_1 = 0\right)$

    H_A: There is a linear relationship between SAT Verbal and Math scores. $\left(\beta_1 \neq 0\right)$

b) **Linearity condition:** The scatterplot is straight enough to try a linear model.
    Independence assumption: The residuals plot is scattered.
    Does the plot thicken? condition: The spread of the residuals is consistent.
    Nearly Normal condition: The histogram of residuals is unimodal and symmetric, with one possible outlier. With the large sample size, it is okay to proceed.

    Since conditions have been satisfied, the sampling distribution of the regression slope can be modeled by a Student's t-model with $(162 - 2) = 160$ degrees of freedom. We will use a regression slope t-test. The equation of the line of best fit for these data points is:
    $$\widehat{Math} = 209.554 + 0.675075(Verbal).$$

    The value of $t = 11.9$. The P-value of less than 0.0001 means that the association we see in the data is unlikely to occur by chance. We reject the null hypothesis, and conclude that there is strong evidence of a linear relationship between SAT Verbal and Math scores. Students with higher SAT-Verbal scores tend to have higher SAT-Math scores.

37. Football salaries.

a) The hypotheses to test the association between the two variables:
    H_0: There is no linear relationship between team salaries and number of wins. $(\beta_1 = 0)$.

    H_A: There is a linear relationship between team salaries and wins. $(\beta_1 \neq 0)$.

b) The t-value shown $= 1.58$ with a P-value $= 0.124$. We fail to reject the null hypothesis. There is no evidence of a linear relationship between team salary and team wins in 2006.

39. Fuel economy, part 2.

   a) 95% confidence interval:

$$b_1 \pm t^*_{n-2} \times SE(b_1) = -8.2136 \pm 2.010*0.6738 = -8.2136 \pm 1.3543 = (-9.57, -6.86)$$

   b) We are 95% confident that the mileage of cars decreases by between 6.86 and 9.57 miles per gallon for each additional 1000 pounds of weight.

41. Wilshire 5000.

   a) The hypotheses to test the association between the two variables:

   H_0: There is no linear association between market return and fund flows. ($\beta_1 = 0$)

   H_A: There is a linear relationship between market return and fund flows. ($\beta_1 \neq 0$)

   b) The t-value shown = 5.75 with a P-value < 0.001. We reject the null hypothesis. There is strong evidence of a linear relationship between money invested in mutual funds and market performance.

   c) Greater investment in mutual funds tends to be associated with higher market return. There are some unusual observations that should be investigated, finding out when and why they occurred.

43. Cost index.

   a) The hypotheses to test the association between the two variables:

   H_0: There is no linear association between the Index in 2006 and 2007. ($\beta_1 = 0$)

   H_A: There is a linear relationship between the Index in 2006 and 2007. ($\beta_1 \neq 0$)

   b) The t-value shown = 8.17 with a P-value < 0.001. We reject the null hypothesis. There is strong evidence of a linear relationship between the Index in 2006 and 2007.

   c) $R^2 = 83.7\%$ indicating a fairly strong linear association. 83.7% of the variation in the Index in 2007 can be explained by the Index in 2006. The association is strong enough to make accurate predictions.

   d) On average, as one increases so does the other. However, this does not mean if the Index is high in one particular year, it is necessarily high in the other year.

45. Oil prices.

   a) The scatterplot shows two distinct clusters of data and perhaps a curved relationship when both clusters are combined which violates the conditions for linear regression.

   b) If there is a relationship between the variables, it does not appear linear. There are unusual patterns in the data – the lower values showing a mostly flat relationship. The higher values of production show much higher values of crude prices in a cluster.

47. Printers. The hypotheses to test the association between the two variables:

H_0: There is no linear association between speed and the cost of a printer. ($\beta_1 = 0$)

H_A: There is a linear relationship between speed and the cost of a printer. ($\beta_1 \neq 0$)

Conditions:

Linearity condition: The scatterplot shown is at least somewhat curved with outliers.
Independence assumption: The residuals plot shows a similar curved pattern with some high outliers.
Equal spread condition: The residuals do not show a uniform spread as values increase.
Nearly Normal condition: The Normal probability plot of residuals is somewhat curved and the histogram of the residuals is skewed to the right.
Because of the violation of conditions, inference is not appropriate.

49. Fuel economy, part 3.

a) The regression equation predicts that cars that weigh 2500 pounds will have a mean fuel efficiency of $48.7393 - 8.21362(2.5) = 28.20525$ miles per gallon.

$$\hat{y}_v \pm t^*_{n-2}\sqrt{SE^2(b_1) \cdot (x_v - \overline{x})^2 + \frac{s_e^2}{n}}$$

$$= 28.20525 \pm (2.014)\sqrt{0.6738^2 \cdot (2.5 - 2.8878)^2 + \frac{2.413^2}{50}}$$

$$\approx (27.34,\ 29.07)$$

We are 95% confident that cars weighing 2500 pounds will have mean fuel efficiency between 27.34 and 29.07 miles per gallon.

b) The regression equation predicts that cars that weigh 3450 pounds will have a mean fuel efficiency of $48.7393 - 8.21362(3.45) = 20.402311$ miles per gallon.

$$\hat{y}_v \pm t^*_{n-2}\sqrt{SE^2(b_1) \cdot (x_v - \overline{x})^2 + \frac{s_e^2}{n} + s_e^2}$$

$$= 20.402311 \pm (2.014)\sqrt{0.6738^2 \cdot (3.45 - 2.8878)^2 + \frac{2.413^2}{50} + 2.413^2}$$

$$\approx (15.44,\ 25.37)$$

We are 95% confident that a car weighing 3450 pounds will have fuel efficiency between 15.44 and 25.37 miles per gallon.

51. Wilshire 5000, part 2.

a) The regression equation predicts that the Fund Flows for a Market Return of 8% is:

$\widehat{FundFlows} = 9599.10 + 1156.40 * 8 = 18,850.30$

$\hat{y}_\nu \pm t^*_{n-2} \times SE$ where

$$SE(\hat{\mu}_\nu) = \sqrt{SE^2(b_1) \times (x_\nu - \overline{x})^2 + \frac{s_e^2}{n} + s_e^2} = \sqrt{201.1^2 \times (8-0.68)^2 + \frac{10999.39^2}{154} + 10999.39^2}$$

$$= \sqrt{40,441.21 \times 53.5754 + 785,627.15 + 120,986,580.37} = 11,132.78$$

The 95% prediction interval:
$\hat{y}_\nu \pm 1.9757 \times 11,132.78 = 18,850.30 \pm 21,995.03 = (-3144.73, 40,845.33)$

b) Since the SE for the slope is relatively large and the R^2 is relatively small (18%), predictions using this regression will be imprecise.

53. All the efficiency money can buy.
   a) We'd like to know if there is a linear association between price and fuel efficiency in cars. We have data on 2004 model year cars, with information on highway MPG and retail price.

   H_0: There is no linear relationship between highway MPG and retail price. ($\beta_1 = 0$)

   H_A: Highway MPG and retail price are linearly associated. ($\beta_1 \neq 0$)

   b) The scatterplot fails the Linearity condition. It shows a bend and it has an outlier. There is also some spreading from right to left, which violates the Equal Spread condition.
   c) Since the conditions are not satisfied, we cannot continue this analysis.

55. Youth unemployment, part 2.
   a) The 95% prediction interval shows the interval of uncertainty for a single predicted male unemployment rate, given a specific female unemployment rate.

   b) The 95% confidence interval shows the interval of uncertainty for the mean male unemployment rate given a sample of female unemployment rates. Because this is an interval for an average, the variation or uncertainty is less, so the interval is narrower than the 95% prediction interval.

   c) The unusual observation is the former Yugoslav Republic of Macedonia. Besides being an outlier, it is also a potential leverage value because its female rate is so removed from the average female rate for this sample of countries. Without this leverage value,

   $\widehat{FundFlows} = 4.4588 + 0.6295^* Female\_Rate$. Note that the slope remains significant (P < 0.001), and the R^2 decreases slightly (0.68) compared to the output in Exercise 9 without this leverage value (76.7%).

57. Energy use, part 2.
   a) The 95% prediction interval shows the interval of uncertainty for the predicted energy use in 2004 based on energy use in 1990 for a single country.

   b) The 95% confidence interval shows the interval of uncertainty for the mean energy use in 2002 based on the same energy use in 1990 for a sample of countries. Because this is an interval for an average, the variation or uncertainty is less, so the interval is narrower than the prediction interval.

59. Seasonal spending.
   a) Conditions:
   Linearity condition: The scatterplot is straight enough to try linear regression.
   Independence assumption: One cardholder's spending should not affect another's spending. The residuals plot shows not pattern. These 99 cardholders are a random sample of cardholders.
   Equal spread condition: The residuals plot shows some increased spread for larger values of December charges.

Nearly Normal condition: The histogram of residuals is unimodal and symmetric with two high outliers.

Scatterplot	Residuals Plot	Residuals Histogram

We should proceed cautiously. There are some issues with the conditions for regression.

The regression model is: $\widehat{January} = 120.73 + 0.6995(December)$

b) The regression equation predicts that cardholders who charged \$2000 in December will charge $120.73 + 0.6995(2000) = \$1519.73$ in January, on average.

c) Cardholders charged an average of \$1336.03 in December.

$$\hat{y}_\nu \pm t^*_{n-2} \sqrt{SE^2(b_1) \cdot (x_\nu - \overline{x})^2 + \frac{s_e^2}{n}}$$

$$= 1519.73 \pm (1.9847)\sqrt{0.0562^2 \cdot (2000 - 1336.03)^2 + \frac{874.5^2}{99}} \approx (1330.24, 1709.24)$$

We are 95% confident that the average January charges for a cardholder that charged \$2000 in December will be between \$1330.24 and \$1709.24.

d) We are 95% confident that a cardholder who charged \$2000 in December charged between \$290.76 and \$669.76 less than \$2000 in January, on average.

e) The residuals show increasing spread, so the confidence intervals may not be valid. We should be very cautious when attempting to interpret them too literally.

61. Environment.

a) H_0: There is no linear relationship between *Population* and *Ozone level*. $(\beta_1 = 0)$

H_A: There is a positive linear relationship between *Population* and *Ozone level*. $(\beta_1 > 0)$

The t-value $= \dfrac{b_1}{SE(Coeff)} = \dfrac{6.650}{1.910} = 3.48$ resulting in a P-value of 0.0018. We reject the null hypothesis. There is strong evidence of a positive linear relationship between the ozone level and the population. Cities with larger populations tend to have higher ozone levels.

b) The city population is a good predictor of ozone level. Population explains 84% (R^2) of the variation in ozone level and s is just over 5 parts per million. This is a high percentage and indicates a strong linear relationship.

Chapter 17 – Understanding Residuals

SECTION 17.1

1. The different segments do not appear to be scattered at random throughout the scatterplot. Rather, it appears that different segments may have different slopes. For instance, the market segment represented in red appears to have a steeper slope than the market segment represented in green. Since the relationship between cardholders' spending in January and spending in December may differ by market segment, using the same model for all market segments would lead to poor predictions.

3. Yes, the fitted regression lines clearly show that the relationship between January and December spending is not the same for all five segments. The slopes are quite different. Therefore, using one overall model to predict January spending based on December spending for all five market segments would be very misleading.

SECTION 17.2

5.

   a) $\hat{y} = 8.10 + 0.913x = 8.10 + 0.913(500) = 464.6$ or $\$464,600$

   b) To find the prediction interval, first find the standard error

   $$SE(\hat{y}_v) = \sqrt{SE^2(b_1) \times (x_v - \bar{x})^2 + \frac{s_e^2}{n} + s_e^2}$$

   $$SE(\hat{y}_v) = \sqrt{(0.0873)^2 \times (500 - 10.4)^2 + \frac{(1.477)^2}{10} + (1.477)^2} = 42.77$$

   The 95% prediction interval is

   $$\hat{y} \pm t^*_{n-2} \times SE = 464.6 \pm 2.306 \times 42.77$$

   $(365.97, 563.23)$ $\$1000$'s or $\$365,970$ to $\$563,230$.

   c) The prediction of sales on a day with 500 employees is not useful. In this case we are extrapolating way beyond the range in the number of employees used to fit the model (2 to 20). Also, it is not likely the bookstore will ever have 500 employees working on a given day.

7.

   a) $\hat{y} = -14,228 + 36.87x = -14,228 + 36.87(10,000) = \$354,472$

   b) Extrapolating this far beyond the range of x values used to fit the model results in unreliable predictions. The largest venues included in the data set have ticket sales of about 3000. Also, not sure if the data on ticket sales used to fit the model were sell out concerts.

SECTION 17.3

9.

   a) With the addition of the point $(100, \$42,000)$, the new regression line predicting *Sales (y)* from *Number of people working (x)* is $\hat{y} = 14.32 + 0.297x$. .

   b) The value of the intercept is higher (14.32 compared to 8.10) and the slope is much lower (0.297 compared to 0.913) from the original estimated regression line.

   c) The point is high leverage because $x = 100$ is far from the mean number of employees working at the bookstore ($\bar{x} = 10.4$). The point is also influential because its presence changes the slope of the regression line.

d) The new regression line predicts that sales will be $44,020 if 100 people are working. The residual, $42,000 - $44,020 = -$2020 is not really that large. This is because the line is so influenced by the new point that it passes close to it.

SECTION 17.5

11.

a)

Quarter	Sales ($M)	Predicted	Residual
1	12	19.02	-7.02
2	41	23.89	17.11
3	15	28.76	-13.76
4	48	33.63	14.37
5	25	38.5	-13.5
6	55	43.37	11.63
7	23	48.24	-25.24
8	69	53.11	15.89
9	51	57.98	-6.98
10	80	62.85	17.15
11	54	67.72	-13.72
12	87	72.59	14.41
13	64	77.46	-13.46
14	94	82.33	11.67
15	62	87.2	-25.2
16	108	92.07	15.93

b)

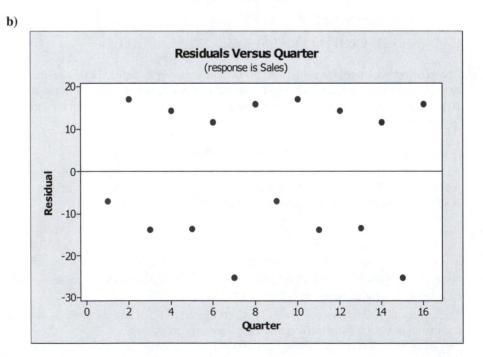

c) $D = \dfrac{\sum(e_t - e_{t-1})^2}{\sum e_t^2} = \dfrac{(17.11-(-7.02))^2 + \ldots + (15.93-(-25.2))^2}{(-7.02)^2 + \ldots + (15.93)^2} = \dfrac{14092.67}{3889.78} = 3.623$

d) Use the standard Dubin-Watson table looking up a sample size of $n=16$, number of predictors $k=1$, and $\alpha = 0.05$ to obtain the critical values $d_L = 1.10$ and $d_U = 1.37$.

e) No. Since $D = 3.623 > d_U = 1.37$ we can conclude that there is no evidence of positive autocorrelation.

f) Yes. Since $D = 3.623 > 4 - d_L = 4 - 1.10 = 2.9$ we can conclude that there is evidence of negative autocorrelation.

13.
a) Consulting the standard Durbin-Watson table with $k = 1$ and $n = 45$, we obtain $d_L = 1.48$ and $d_U = 1.57$. The observed value $D=0.51$ is less than d_L therefore we can reject the null hypothesis of no autocorrelation. There is evidence of positive autocorrelation.

b) The statistic tests for time-related correlation.   However, these data are not time series data. Indicates a pattern in the residuals not captured by the model.

SECTION 17.6

15. No. The scatterplot shows a linear relationship and the residual plot has no obvious pattern, therefore no transformation is recommended.

SECTION 17.7

17. A re-expression is called for since the relationship between *Cost per Unit* and *Number of Units* is curvilinear. Try taking the log or reciprocal of *Cost per Unit* to make the relationship more nearly linear.

19.
a) An appropriate re-expression for the number of customers is to take the Log.

b) The Log transformation corresponds to the 0 power in the ladder of powers.

CHAPTER EXERCISES

21. **Marriage age 2007.**
a) The trend in age at first marriage for American women is very strong over the entire time period recorded on the graph, but the direction and form are different for different time periods. The trend appears to be somewhat linear, and consistent at around 22 years, up until about 1940, when the age seemed to drop dramatically, to under 21. There is a negative trend from 1945-1960. After that period until about 1970, the trend appears non-linear and slightly positive. From 1975 to about 1995, the trend again appears linear and positive. The marriage age rose rapidly during this time period.

b) The association between age at first marriage for American women and year is strong over the entire time period recorded on the graph, but some time periods have stronger trends than others.

c) The correlation, or the measure of the degree of linear association is not high for this trend. The graph, as a whole, is non-linear. However, certain time periods, like 1975 to 1995, have a high correlation.

d) Overall, the linear model is not appropriate. The scatterplot is not "straight enough" to satisfy the condition. You could fit a linear model to the time period from 1975 to about 1995, but this seems unnecessary. The ages for each year are reported, and, given the fluctuations in the past, extrapolation seems risky.

23. Human Development Index.
   a) Fitting a linear model to the association between HDI and GDPPC would be misleading, since the relationship is not straight.

   b) If you fit a linear model to these data, the residuals plot will be curved downward.

   c) Setting aside the single data point corresponding to Luxembourg will not improve the model. The relationship will still be curved.

25. Good model?
   a) The student's reasoning is not correct. A scattered residuals plot, not high R^2, is the indicator of an appropriate model. Once the model is deemed appropriate, R^2 is used as a measure of the strength of the model.

   b) The model may not allow the student to make accurate predictions. The data may be curved, in which case the linear model would not fit well. The consultant would need to see some graphical displays to decide.

27. Movie dramas.
   a) The units for the slopes of these lines are millions of dollars per minutes of running time.

   b) The slopes of the regression lines are the same. Dramas and movies from other genres have costs for longer movies that increase at the same rate.

   c) The regression line for dramas has a lower y-intercept. Regardless of running time, dramas cost about 20 million dollars less than other genres of movies of the same running time.

29. Oakland passengers.
   a) According to the linear model, the use of the Oakland airport has been increasing by about 59,700 passengers per year, starting at about 283,000 passengers in 1990.

   b) About 71% of the variability in the number of passengers can be accounted for by the model.

   c) Errors in prediction based on the model have a standard deviation of 104,330 passengers.

   d) From technology the Durbin-Watson statistic = 0.911 with a P-value < 0.001. There is strong evidence of autocorrelation.

   e) No, the model would not be useful in predicting the number of passengers in 2010. This year would be an extrapolation too far from the years we have observed.

   f) The negative residual is September of 2001. Air traffic was artificially low following the attacks on 9/11/2001.

31. Unusual points.
   a) 1) The point has high leverage and a small residual.
      2) The point is not influential. It has the potential to be influential, because its position far from the mean of the explanatory variable gives it high leverage. However, the point is not exerting much influence, because it reinforces the association.
      3) If the point were removed, the correlation would become weaker. The point heavily reinforces the positive association. Removing it would weaken the association.
      4) The slope would remain roughly the same, since the point is not influential.

   b) 1) The point has high leverage and probably has a small residual.
      2) The point is influential. The point alone gives the scatterplot the appearance of an overall negative direction, when the points are actually fairly scattered.

3) If the point were removed, the correlation would become weaker. Without the point, there would be very little evidence of linear association.
4) The slope would increase, from a negative slope to a slope near 0. Without the point, the slope of the regression line would be nearly flat.

c) **1)** The point has moderate leverage and a large residual.
2) The point is somewhat influential. It is well away from the mean of the explanatory variable, and has enough leverage to change the slope of the regression line, but only slightly.
3) If the point were removed, the correlation would become stronger. Without the point, the positive association would be reinforced.
4) The slope would increase slightly, becoming steeper after the removal of the point. The regression line would follow the general cloud of points more closely.

d) **1)** The point has little leverage and a large residual.
2) The point is not influential. It is very close to the mean of the explanatory variable, and the regression line is anchored at the point (x,y), and would only pivot if it were possible to minimize the sum of the squared residuals. No amount of pivoting will reduce the residual for the stray point, so the slope would not change.
3) If the point were removed, the correlation would become slightly stronger (decreasing to becoming more negative). The point detracts from the overall pattern, and its removal would reinforce the association.
4) The slope would remain roughly the same. Since the point is not influential, its removal would not affect the slope.

33. The extra point.
1) Point **e** is very influential. Its addition will give the appearance of a strong, negative correlation like $r = -0.90$.
2) Point **d** is influential (but not as influential as point e). Its addition will give the appearance of a weaker, negative correlation like $r = -0.40$.
3) Point **c** is directly below the middle of the group of points. Its position is directly below the mean of the explanatory variable. It has no influence. Its addition will leave the correlation the same, $r = 0.00$.
4) Point **b** is almost in the center of the group of points, but not quite. Its addition will give the appearance of a very slight positive correlation like $r = 0.05$.
5) Point **a** is very influential. Its addition will give the appearance of a strong, positive correlation like $r = 0.75$.

35. What's the cause?
1) High blood pressure may cause high body fat.
2) High body fat may cause high blood pressure.
3) Both high blood pressure and high body fat may be caused by a lurking variable, such as a genetic or lifestyle trait.

37. Heating cost.
a) The model predicts a decrease in $2.13 in heating cost for an increase in temperature of 1° Fahrenheit. Generally, warmer months are associated with lower heating costs.

b) When the temperature is 0° Fahrenheit, the model predicts a monthly heating cost of $133.

c) When the temperature is around 32° Fahrenheit, the predictions are generally too high. The residuals (observed – predicted) are negative, indicating that the actual values are lower than the predicted values.

d)
$$\hat{C} = 133 - 2.13(Temp)$$
$$\hat{C} = 133 - 2.13(10)$$
$$\hat{C} = \$111.70$$

According to the model, the heating cost in a month with average daily temperature 10° Fahrenheit is expected to be $111.70.

e) The residual for a 10° day is approximately –$6, meaning that the actual cost was $6 less than predicted, or $111.70 – $6 = $105.70.

f) The model is not appropriate. The residuals plot shows a definite curved pattern. The association between monthly heating cost and average daily temperature is not linear.

g) A change of scale from Fahrenheit to Celsius would not affect the relationship. Associations between quantitative variables are the same, no matter what the units.

39. Interest rates.

a) $r = \sqrt{R^2} = \sqrt{0.774} = 0.88$. The correlation between rate and year is +0.88, since the scatterplot shows a positive association.

b) According to the model, interest rates during this period increased at about 0.25% per year, starting from an interest rate of about 0.64% in 1950.

c) The linear regression equation predicting interest rate from year is

$$\widehat{Rate} = 0.640282 + 0.247637(Year - 1950)$$

$$\widehat{Rate} = 0.640282 + 0.247637(50)$$

$$\widehat{Rate} = 13.022$$

According to the model, the interest rate is predicted to be about 13% in the year 2000.

d) Durbin-Watson statistic = 0.953. P = 0.002 and there is evidence of strong auto-correlation.

e) This prediction is not likely to have been a good one. Extrapolating 20 years beyond the final year in the data would be risky, and unlikely to be accurate.

41. Interest rates, part 2.

a) The values of R^2 are approximately the same, so the models fit comparably well, but they have very different slopes.

b) The model that predicts the interest rate on 3-month Treasury bills from the number of years since 1950 is $\widehat{Rate} = 21.0688 - 0.356578(Year - 1950)$. This model predicts the interest rate to be 3.24%, a rate much lower than the prediction from the other model.

c) We can trust the new predicted value because it is in the middle of the data used for the regression.

d) The best answer is "I can't predict that." 2020 is way beyond the data points given.

43. Colorblind. Sex and Colorblindness are both categorical variables, not quantitative. Correlation is meaningless for them, but we can say that the variables are associated.

45. Residuals.
a) No re-expression needed.

b) Re-express to straighten the relationship.

c) Re-express to equalize the spread.

47. Oakland passengers, part 2.
a) There's an annual pattern, so the residuals cycle up and down.

b) No, this kind of pattern can't be helped by re-expression.

49. Models.

   a) $\hat{y} = 1.2 + 0.8(2) = 2.8$

   b) $\ln \hat{y} = 1.2 + 0.8(2) = 2.8$ so $\hat{y} = 16.44$ (calculate $e^{\hat{y}} = e^{2.8}$)

   c) $\sqrt{\hat{y}} = 1.2 + 0.8(2) = 2.8$ so $\hat{y} = 2.8^2 = 7.84$

   d) $\dfrac{1}{\hat{y}} = 1.2 + 0.8(2) = 2.8$ so $\hat{y} = \dfrac{1}{2.8} = 0.36$

   e) $\hat{y} = 1.2(2)^{0.8} = 2.09$

51. Models, again.

   a) $\hat{y} = 2 + 0.8(\ln 10) = 3.84$

   b) $\log \hat{y} = 5 - 0.23(10) = 2.7$ so $\hat{y} = 10^{2.7} = 501.19$

   c) $\dfrac{1}{\sqrt{\hat{y}}} = 17.1 - 1.66(10) = 0.5$ so $\hat{y} = \left(\dfrac{1}{0.5}\right)^2 = 4$

53. Lobster industry 2008.

   a) Linearity condition violated. Possibly increased variation at higher values. Independence may be in doubt since this is a time series.

   b) This plot looks both straighter and more consistent in variation.

   c) The residuals show a pattern, but it tracks up and down and could not be improved by an alternative transformation.

55. Lobster value.

   a) $\widehat{LogValue} = 6.43619 + 0.000156\, Fisher = 6.43619 + 0.000156(10,000) = 7.99619$ so the *Value* = $10^8 = \$100,000,000$.

   b) *LogValue* grows at 0.000156 per licensed *Fisher*. Possibly more fishers cause a more valuable harvest, but greater value would likely attract more fishers, so the causation might run the other way. Or improving technology could both lead to more value and attract more fishers.

57. GDP.

   a) Although more than 97% of the variation in GDP can be accounted for by this model, we should examine a scatterplot of the residuals to see if it's appropriate.

   b) No. The residuals show clear curvature.

59. Logs (not logarithms).

    a)   The transformation that makes this relationship linear is: $\widehat{\sqrt{BoardFeet}} = -4 + Diameter$
          The model is exact.

    b) $\widehat{\sqrt{BoardFeet}} = -4 + 10 = 6$ so $BoardFeet = 6^2 = 36$

    c) $\widehat{\sqrt{BoardFeet}} = -4 + 36 = 32$ so $BoardFeet = 32^2 = 1024$

61. OECD GDP.

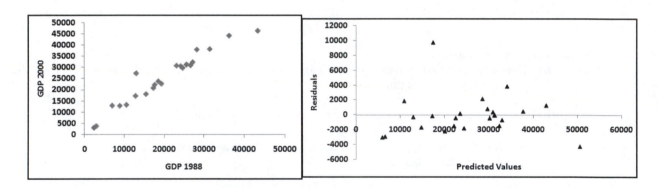

Regression Summary Output					
R	0.9705				
R Squared	0.94187025				
Sum Res Squared	172861656				
Standard Error	2803.09615				
Observations	24				
				P value	
	Coefficient	Std Error	t Stat	1-Sided	2-Sided
Intercept	3295.79525	1306.837	2.521964	0.009706	0.019412
GDP 1988	1.09016126	0.057746	18.87862	2.22E-15	4.43E-15

The regression has a high R^2 but the residuals show two influential points, Ireland with the largest positive residual and Switzerland with the largest negative residual. Assuming the data for these countries are correct, further work needs to be done to understand why each is unusual. Reports both with and without the outliers should be compared.

63. Human Development Index, part 3. The histogram is much more unimodal and symmetric after we take logs.

65. Lobster fishers. The plot cannot be straightened by a transformation because it is not a simple monotonically increasing or decreasing shape, but rather, it oscillates up and down.

Chapter 18 – Multiple Regression

SECTION 18.1

1.

   a) $\hat{y} = 20{,}986.09 - 7483.10(2) + 93.84(1000) = \$99859.89$.

   b)   The residual for the house that just sold for \$135,000 is \$135,000 - \$99,859.89 = \$35,140.11.

   c)   The house sold for more than the model predicted. The model underestimated the selling price.

SECTION 18.2

3.

   a)   *USGross* $= -22.9898 + 1.13442 Budget + 24.9724 Stars - 0.403296 RunTime$

   b)   After allowing for the effects of *Stars* and *RunTime*, each additional million dollars in the budget for making the film yields about 1.13 million dollars in gross revenue.

SECTION 18.3

5.

   a.   **Linearity**: The scatterplot shows that the relationship is reasonably linear; no curvature is evident in the scatterplot. The linearity condition is satisfied.

   b.   **Equal Spread:** The scattering of points seems to increase as the budget increases. The points are much more scattered (spread apart) to the right than to the left. Consequently, it appears that the equal spread condition is not satisfied.

   c.   **Normality**: A scatterplot of two variables gives us no information about the distribution of the residuals. Therefore we cannot determine if the normality assumption is satisfied.

SECTION 18.4

7.

   a)   The null hypothesis for testing the coefficient associated with *Stars* is: H_0: $\beta_{Stars} = 0$.

   b)   The *t*-statistic for this test is given by:

$$t_{n-k-1} = \frac{b_j - 0}{SE(b_j)}$$

$$t_{120-3-1} = \frac{24.9724}{5.884} = 4.244$$

   c)   The associated P-value is ≤ 0.0001.

   d)   We can reject the null hypothesis. There is sufficient evidence to suggest that the coefficient of *Stars* is not equal to zero. *Stars* is a significant predictor of a film's gross income.

SECTION 18.5

9.

   a)   For this regression equation, $R^2 = 0.4192977$ or 41.93%. This tells us that 41.93% of the variation in the dependent variable *USGross* is explained by the regression model that include s the predictors *Budget, Run Time* and *Stars*.

   b)   The *Adjusted* R^2 value is slightly lower than the value of R^2. This is because the *Adjusted* R^2 is adjusted downward (imposing a "penalty") for each new predictor variable added to the model. Consequently, it allows the comparison of models with different numbers of predictor variables.

CHAPTER EXERCISES

11. Police salaries.

a) **Linearity condition**: The scatterplots appear at least somewhat linear but there is a lot of scatter.
Randomization condition (Independence Assumption): States may not be a random sample but may be independent of each other.
Equal spread condition (Equal Variance Assumption): The scatterplot of Violent Crime vs Police Officer Wage looks less spread to the right but may just have fewer data points. Residual plots are not provided to analyze.
Nearly Normal condition (Normality Assumption): To check this condition, we will need to look at the residuals which are not provided for this example.

b) The R^2 of that regression would be $(0.103)^2 = 0.011 = 1.1\%$.

13. Police salaries, part 2.

a) The regression model: $\widehat{Violent\ Crime} = 1390.83 + 9.33\,Police\ Officer\ Wage - 16.64\,Graduation\ Rate$

b) After allowing for the effects of *Graduation Rate*, states with higher *Police Officer Wages* have more *Violent Crime* at the rate of 9.33 crimes per 100,000 for each dollar per hour of average wage.

c) $\widehat{Violent\ Crime} = 1390.83 + 9.33*20 - 16.64*70 = 412.63$ crimes per 100,000.

d) The prediction is not very good. The R^2 of that regression is only 53.0%.

15. Police salaries, part 3.

a) The $2.26 = \dfrac{9.33}{4.125}$

b) There are 40 states used in this model. The degrees of freedom are shown to be 37, which is equal to $n - k - 1 = 40 - 2 - 1 = 37$. There are two predictors.

c) The *t*-ratio is negative because the coefficient is negative meaning that *Graduation Rate* contributes negatively to the regression. After allowing for the effects of other factors, an increase of one point in *Graduation Rate* is associated with a decrease on average of about 16.64 in Police Salary.

17. Police salaries, part 4.

a) The hypotheses are: $H_0 : \beta_{Officer} = 0$; $H_A : \beta_{Officer} \neq 0$.

b) $P = 0.0297$ which is small enough to reject the null hypothesis at $\alpha = 0.05$ and conclude that the coefficient is different from zero.

c) The coefficient of *Police Officer Wage* reports the relationship after allowing for the effects of *Graduation Rate*. The scatterplot and correlation were only concerned with the relationship between *Police Officer Wages* and *Violent Crime* (two variables).

19. Police salaries, part 5. This is a causal interpretation, which is not supported by regression. For example, among states with high graduation rates, it may be that those with higher violent crime rates spend more to hire police officers, or states with higher costs of living must pay more to attract qualified police officers but also have higher crime rates.

21. Police salaries, part 6.
    Constant Variance Condition (Equal Spread): met by the residuals vs. predicted plot.
    Nearly Normal Condition: met by the Normal probability plot.

23. Real estate prices.
   a) Incorrect: Doesn't mention other predictors; suggests direct relationship between only two variables: *Age* and *Price*.

   b) Correct

   c) Incorrect: Can't predict *x* from *y*

   d) Incorrect interpretation of R^2 (this model accounts for 92% of the of the variability in *Price*)

25. Appliance sales.
   a) Incorrect: This is likely to be extrapolation since it is unlikely that they observed any data points with no advertising of any kind.

   b) Incorrect: Suggests a perfect relationship

   c) Incorrect: Can't predict one explanatory variable (*x*) from another

   d) Correct

27. Cost of pollution.
   a) The negative sign of the coefficient for ln(number of employees) means that for businesses that have the same amount of sales, those with more employees spend less per employee on pollution abatement on average. The sign of the coefficient for ln(sales) is positive. This means that for businesses with the same number of employees, those with larger sales spend more on pollution abatement on average.

   b) The logarithms mean that the effects become less severe (in dollar terms) as companies get larger either in *Sales* or in *Number of Employees*.

29. Home prices.
   a) $\widehat{Price} = -152,037 + 9530 Baths + 139.87 Area$

   b) $R^2 = 71.1\%$

   c) For houses with the same number of bathrooms, each square foot of area is associated with an increase of $139.87 in the price of the house, on average.

   d) The regression model says that for houses of the same size, there is no evidence that those with more bathrooms are priced higher. It says nothing about what would actually happen if a bathroom were added to a house.

31. Secretary performance.
   a) The regression equation:
$\widehat{Salary} = 9.788 + 0.110 Service + 0.053 Education + 0.071 Test Score + 0.004 Typing\ wpm + 0.065 Dictation\ wpm$

   b) $\widehat{Salary} = 9.788 + 0.110*120 + 0.053*9 + 0.071*50 + 0.004*60 + 0.065*30 = 29.205$ or $29,205

   c) The *t*-value is 0.013 with 24 *df* and a P-value = 0.9897 (two-tailed), which is not significant at $\alpha = 0.05$.

    d)   You could take out the explanatory variable X4 (*typing speed*) since it is not significant.

    e)   *Age* is likely to be collinear with several of the other predictors already in the model. For example, secretaries with longer terms of *Service* will naturally also be older.

33. Wilshire 5000.

    a)   After allowing for the effects measured by the Wilshire 5000 index, an increase of one point in the *Unemployment Rate* is associated with a decrease on average of about \$3719.55 million in the *Funds flow*.

    b)   The *t*-ratio divides the coefficient by its standard error. The coefficient here is negative.

    c)   The hypotheses are: $H_0 : \beta_{unemp\_rate} = 0$; $H_A : \beta_{unemmp\_rate} \neq 0$; P-value < 0.0001 is very small, so reject the null hypothesis.

35. Lobster industry 2008, part 2.

    a)   *Estimated* $\Pr ice / lb = 0.565367 + 1.30744 Traps (M) - 0.00012 Fishers - 9.97788 Catch / Trap$

    b)   Residuals show no pattern and have equal spread. The histogram of residuals is nearly Normal. There is a question whether values from year to year are mutually independent.

    c)   The hypotheses are: $H_0 : \beta_{catch/trap} = 0$; $H_A : \beta_{catch/trap} \neq 0$; P-value $= 0.3781$. Consequently, we fail to reject the null hypothesis and conclude that *Catch/Trap* is not a significant predictor of *Price/lb*; it does not contribute to the model.

    d)   After allowing for other variables, it does make sense that when the *Catch/Trap* decreases, the *Price* will increase.

    e)   The adjusted R^2 accounts for the number of predictors. These models have different numbers of predictors, so this is an appropriate comparison. The previous model in Chapter 17 had a similar value for R^2. It makes sense that the model accounts for almost 95% of the variability in Price – as the supply decreases (in the form of # of *Fishers, Catch/Trap*), the *Price* increases.

37. Wal-Mart revenue, part 2.

    a) $\widehat{PBE} = 87.0 - 0.345 CPI + 0.000011 Personal\ Consumption + 0.0001 Retail\ Sales$

	Coefficients	Standard Error	t Stat	P-value
Intercept	87.00892605	33.59897163	2.589631	0.013908
CPI	-0.344795233	0.120335014	-2.86529	0.007002
Personal Consumption	1.10842E-05	4.40271E-06	2.51759	0.016546
Retail Sales Index	0.000103152	1.54563E-05	6.67378	1.01E-07

Regression Statistics	
Multiple R	0.816425064
R Square	0.666549886
Adjusted R Square	0.637968448
Standard Error	2.326701861
Observations	39

    b) $R^2 = 66.7\%$ and all *t*-ratios are significant. It looks like these variables can account for much of the variation in Wal-Mart revenue.

39. **Right-to-work laws.**

   a) $Logit(Right-to-work) = 6.19951 - 0.106155\,publ - 0.222957\,pvt$

   b)   To find the predicted log odds (logit) of the probability that a state has a right-to-work law if 20% of the public and 10% of the private sector employees are in unions, set *publ* = 20 and *pvt* = 10 in the estimated regression equation: $6.19951 - 0.106155\,(20) - 0.222957(10) = 1.847$.

   c)   The predicted probability of that state is

   $$\hat{p} = \frac{1}{1+e^{-(1.847)}} = \frac{1}{1+0.1577095} = 0.864$$

   d)   To find the predicted log odds (logit) of the probability that a state has a right-to-work law if 40% of the public and 20% of the private sector employees are in unions, set *publ* = 40 and *pvt* = 20 in the estimated regression equation: $6.19951 - 0.106155\,(40) - 0.222957(20) = -2.506$.

   e)   The associated predicted probability is

   $$\hat{p} = \frac{1}{1+e^{-(-2.506)}} = \frac{1}{1+12.253725} = 0.075$$

41. **Motorcycles.**
   The scatterplot of *MSRP* versus *Wheelbase* indicates that this relationship is not linear. While there is a positive relationship between *Wheelbase* and *MSRP*, a curved pattern is evident. The scatterplot of *MSRP* versus *Displacement* indicates that this relationship is positive and linear. The scatterplot of *MSRP* versus *Bore* indicates that this relationship is also positive and linear. Based on these plots, it appears that both *Displacement* and *Bore* would be better predictors of *MSRP* than *Wheelbase*.

43. **Motorcycles, part 3.**
   a)   Yes, with an $R^2 = 90.9\%$ says that most of the variability of *MSRP* is accounted for by this model.

   b)   No, in a regression model, you can't predict an explanatory variable from the response variable.

45. **Burger King nutrition.**

   a)   With an $R^2 = 100\%$, the model should make excellent predictions.

   b)   The value of *s*, 3.140 calories, is very small compared to the initial standard variation of *calories*. This means that the model fits the data quite well, leaving very little variation unaccounted for.

   c)   No, the residuals are not all 0. Indeed, we know that their standard deviation is *s* = 3.140 calories. They are very small compared with the original values. The true value of R^2 was rounded up to 100%.

Chapter 19 – Building Multiple Regression Models

SECTION 19.1

1.

    a) *Unionization Status* = 1 if company is unionized.

          Unionization Status = 0 if not.

    b) *Gender* = 1 if Female.

          Gender = 0 if Male.

    c) *Account Status* = 1 if paid on time.

          Account Status = 0 if past due.

    d) Use two dummy variables to indicate three political party affiliations. For example, let

          Democrat = 1 if the individual is registered Democrat, 0 otherwise.

          Republican = 1 if the individual is registered Republican, 0 otherwise.

3.

    a) This plot shows parallel or nearly parallel lines for Comedies and Action films. This means that the slopes of the relationships between *USGross* vs. *Budget* are very similar for both types of films but that the intercepts are different. This suggests that the use of an indicator variable to categorize the films into one or the other group would be appropriate.

    b) Use one dummy variable as follows: Let *Comedy* = 1 if the film is a Comedy and *Comedy* = 0 if the film is an Action picture.

SECTION 19.2

5.

    a) Since there are four different types of residence, three dummy variables are needed. These can be coded as follows: Let

          Apartment = 1 if the residence is an apartment, 0 otherwise.

          Condo = 1 if the residence is a condo, 0 otherwise.

          Townhouse = 1 if the residence is a townhouse, 0 otherwise.

    b) Since there are three different possibilities for employment status, two dummy variables are needed. These can be coded as follows: Let

          Full Time = 1 if employment status is full-time, 0 otherwise.

          Part Time = 1 if employment status is part-time, 0 otherwise.

7.

    a) This plot shows that the lines for Comedies and Adventure films are not parallel. This means that the slopes of the relationships between *USGross* vs. *Budget* are not the same for both types of films. Consequently, the use of just an indicator variable to categorize the films into one or the other group would not be enough to account for the different slopes. It will be necessary to include an interaction term as well.

    b) Use one dummy variable as follows: Let *Comedy* = 1 if the film is a Comedy and *Comedy* = 0 if the film is an Adventure film (coding in this way makes Adventure the base level).

    c) The interaction term is constructed as the product of *Budget* and the indicator variable *Comedy*.

SECTION 19.3

9.

Leverage is defined as the amount of change in $\hat{y}$ predicted by the regression recomputed with a unit change in the y – value for a given case. Hence, if the movie *King Kong* had earned \$1M more than it did (this is the change in the actual y –value) and the resulting leverage was 0.189, this means that the model's prediction of *King Kong's* U.S. gross revenue would change by \$ 0.189 M.

SECTION 19.4

   11.

   The correlation matrix shows some very high correlations among the potential independent variables. Specifically, there is a very high correlation between *Income* and *Net Worth*, *Income* and *Age*, and *Net Worth* and *Age*. Depending on the order in which the predictors are chosen for entry by the stepwise regression search, it is possible that the resulting model will not be the best as determined by the "all subsets" approach.

SECTION 19.5

   13.

   Since the output shows the results of regressing *Age* on the other predictor variables, the VIF for *Age* is

$$VIF_j = \frac{1}{1 - R_j^2} = \frac{1}{1 - .9875} = 80$$

SECTION 19.6

   15.

   a) The scatterplot indicates that the relationship between the *Price* and *Age* of houses is curved upward at both extremes (low and high ends) of age. This tells us that the most expensive homes are either quite new or very old.

   b) The reason why the slope for *Age* is not statistically significant in a linear regression of *Price* on *Age* is because the relationship between these two variables is not linear. However, as seen in the scatterplot, *Price* is related to *Age* in a nonlinear fashion. Consequently, *Age* should not be dropped from the model. Rather the model should be modified to capture the nonlinear, curved relationship.

   c) A recommended option is to add a quadratic term to the model (add the variable Age^2).

CHAPTER EXERCISES

   17. Pizza ratings.
   a) Cheese pizzas (Type 1) scored about 15.6 points higher, after allowing for the effects of *calories* and *fat*. Cheese pizzas can be expected to sell better than pepperoni pizzas based on these results.

   b) We should check the scatterplots of the data to check for linearity. Check the plot of the residuals against predicted values or against each predictor, looking for pattern or outliers. We should also check for evidence that the residuals are nearly Normal with a Normal probability plot or histogram.

   19. Pizza ratings, part 2.
   a)   Both Reggio's and Michelina's pizzas are predicted to have high scores but both received lower scores (more than 30 points lower) than we would otherwise have predicted from the model.

   b) Yes, their unusually large predicted values indicate that they must be different from the other pizzas in some ways, and their large (negative) residuals indicate that they would be influential and affect the regression by "pulling" the regression line downward. The values are much lower than the other values in the dataset and will affect the slope of the regression. The result will be that the large negative residual values will make the regression line less positive. We could calculate the *t*-ratio for the two types of pizza – Michelina's and Reggio's. If the *t*-values are large, we can reject the null hypothesis that either pizza fits the regression model and determine that either value is an outlier. Other supporting calculations would be the leverage values and Cook's distances.

21. Wal-Mart revenue.
a)   An indicator or "dummy" variable.

b)   Sales are about 10.5 billion dollars higher in December, after accounting for the CPI.

c)   Must assume that the slope is the same for the December points as for the others. That appears to be true in the scatterplot.

23. Pizza ratings, part 3.
a)   There are different slopes for the two types.

b)   This is an interaction term. It says that cheese pizzas (Type 1) have scores that grow less rapidly with Calories (a slope 0.46 smaller) than do pepperoni pizzas. The slope of taste score on calories, after allowing for the linear effects of fat and removing the influence of the two outlying pizzas, is estimated to be 1.92 points per gram for pepperoni pizzas and $1.92 - 0.4615 = 1.45$ points per gram for cheese pizzas.

c)   The adjusted R^2 for this model is higher than the adjusted R^2 for the previous model. Also, the t-ratios are larger. Overall, this looks like a more successful regression model. We have identified a consistent difference between pepperoni and cheese pizzas and incorporated it into the model. All coefficients are significantly different from zero.

25. Insurance (Life expectancy).
a)   Large Cook's distance suggests that Alaska is influential. Its leverage is high, and its residual is one of the lowest.

b)   After allowing for the other predictors, Alaska's life expectancy is 2.23 years lower than expected. The P-value of 0.02 says that Alaska is an outlier.

c)   Each of the models has at least one coefficient with a large P-value (*Income* and *Illiteracy* in both models although smaller values in the dummy variable model). These predictor variables could be omitted to simplify the model for further analysis. The model where Alaska is removed and replaced with a dummy variable model is a better choice because of a higher adjusted R^2 and smaller standard deviation of residuals.

27. Lobster industry 2008, value diagnosed. The *Value* for 1994 is not particularly low, but apparently, judging from its Studentized residuals, it was unexpectedly low for the number of traps and fishers and the water temperature. Any point with a large Cook's distance is potentially influential, so we might want to set 1994 aside to see whether that changes the model.

29. OECD economic regulation (model building).
a)   Adjusted R^2 accounts for the number of predictors. If removing a predictor causes it to increase, then that predictor contributed little to the model. *Primary Education* can be omitted from the model.

b)   Whether the t-ratio for *OECD Regulation* is technically less than 0.05 or greater than 0.05 depends on including *Primary Education* in the model. We might be suspicious that the author, in his desire to claim that regulation harms *GDP/Capita*, has found an irrelevant predictor that has the effect of nudging the P-value just below 0.05.

31. Airlines on time.
a)   With a P-value of 0.686, we would not conclude that the true coefficient is not zero. It cannot be interpreted.

b)   The R^2 of the second regression shows that *Diverted* is linearly related to the other predictors and thus suffers from collinearity.

c)   The same answer as b)

d) $VIF_j = \dfrac{1}{1 - R_j^2} = \dfrac{1}{1 - 0.856} = 6.94$

33. Ticket prices, diagnosed.

   a)   The points may be influential. The points have both unusually low predicted values and large residuals, so this seems likely.

   b)   Two of the coefficients have not changed very much, but the coefficient of #*Shows* has changed from negative but not significant to positive and strongly significant. A point is influential if removing it from the model changes any coefficients in important and influential ways.

   c)   The second model would be better. The first regression was influenced by the two possibly influential points which turned out not to be representative weeks on Broadway due to the strike.

35. Lobster industry 2008, catch model.

   a)   The linearity condition is violated. A re-expression won't help because the curve goes down and then back up rather than consistently rising or falling.

   b)   Using the regression model:

$Catch / Fisher = 2.82022 - 2.5153(3) + 0.898526(3)^2 = 3.361054$

$3.361054 \times 7327 \, fishers = 24,626.44 tonnes$

   c)   *Traps* and *Traps^2* are highly correlated. The collinearity accounts for the change in the coefficient of *Traps*.

Chapter 20 – Time Series Analysis

SECTION 20.1

1.
   a) Yes
   b) No, these data are not recorded sequentially over time. Unemployment data is provided for one month only (August, 2010) and by educational level.
   c) No, while these data represent time (time spent in training by each worker), these data are not recorded sequentially over time. The time ordering of the data is not an important aspect.
   d) Yes

SECTION 20.2

3.
   a)

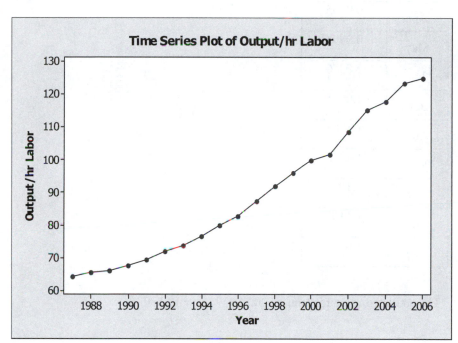

   b) The trend is positive, strong and slightly curved.
   c) No. As expected with yearly data, there is no evidence of a seasonal component.

SECTION 20.3

5.
   a) The 2-point moving averages for the first year (2006) of *Apple Prices* are shown in the table below. Each is found using the following expression:

$$\tilde{y}_t = \frac{\sum\limits_{i=t-L+1}^{t} y_i}{L}$$

For example, for February we have

$$\tilde{y}_2 = \frac{\sum\limits_{i=1}^{2} y_i}{2} = \frac{0.963 + 0.977}{2} = 0.97$$

Month	MA(2)
Jan	*
Feb	0.97
Mar	0.9560
Apr	0.9465
May	0.9895
Jun	1.0370
Jul	1.0995
Aug	1.1905
Sep	1.2455
Oct	1.1970
Nov	1.1135
Dec	1.0580

b) The predicted value for January 2007 is $\tilde{y}_{Dec} = 1.0580$.

SECTION 20.4

7. The Absolute Percentage Error of the forecast for January 2007 is

$$APE = 100 \times \frac{\left| y_t - \hat{y}_t \right|}{\left| y_t \right|}$$

$$APE = 100 \times \frac{\left| 1.034 - 1.058 \right|}{\left| 1.034 \right|} = 2.32\%$$

SECTION 20.5

9. The lag1 version of *Apple Prices* is shown in the table below.

Month	Price$_{lag1}$
Jan	*
Feb	0.963
Mar	0.977
Apr	0.935
May	0.958
Jun	1.021
Jul	1.053
Aug	1.146
Sep	1.235
Oct	1.256
Nov	1.138
Dec	1.089

11. The predicted *Apple Price* for January 2007 is

$$\hat{y}_t = 0.207647 + 1.32993 y_{t-1} - 0.504445 y_{t-2}$$

$$\hat{y}_{Jan2007} = 0.207647 + 1.32993 \times 1.027 - 0.504445 \times 1.089$$

$$\hat{y}_{Jan2007} = 1.024$$

SECTION 20.6

13.

- a) The variable *Jan* is an indicator or dummy variable.
- b) The number of indicator (or dummy) variables included in the model is one less than the number of categories being represented. Since in this model the indicator variables are representing 12 months, then there are 11 indicator variables in the model. Consequently, December is coded as all zeros and represents the base level. This is done to avoid the problem of collinearity.
- c) Referring to the times series plot for four years of *Apple Prices* seasonality and an upward trend is apparent. It is often difficult to discern whether an upward trend evident during a relatively short period is actually a cyclical component tied to the economy. There is no definitive evidence of a cyclical component in the time series plot.

CHAPTER EXERCISES

15. Concepts.

a) The 200-day moving average will be smoother – the moving average with the longer length is smoother.

b) The SES model using $\alpha = 0.10$ is smoother – the smoothing coefficient closer to zero is smoother.

c) When the α is raised from 0.10 to 0.80, the most recent data point is weighted more heavily, so the model responds more quickly and is not as smooth.

17. More concepts.

a) Random walk (rapid and sudden changes in direction) is best modeled using a naïve forecast. Another possibility is the AR(1) model based on the average of prior values in the series.

b) Exponential model for a consistent annual percentage increase.

c) Seasonal dummy variable model, or an AR model that uses sales from the same season in prior years (lag4).

19. Liquid assets.

a)   22,988 $M

Year	Liquid Assets ($M)	3-Yr Avg
2002	18,546	
2003	22,364	
2004	22,413	21,108
2005	19,881	21,553
2006	26,670	22,988

b) From technology, the predicted value for 2007 using single exponential smoothing with smoothing parameter α = 0.2: 22,394 $M

```
Time     C1    Smooth   Predict      Error
1      18546  21289.0  21974.8   -3428.80
2      22364  21504.0  21289.0    1074.96
3      22413  21685.8  21504.0     908.97
4      19881  21324.9  21685.8   -1804.83
5      26670  22393.9  21324.9    5345.14


Forecasts

Period  Forecast    Lower    Upper
6        22393.9  16238.3  28549.5
```

21. Banana prices.

a)   475.88 $/tonne

	$/tonnes	3-Yr Avg
2000	422.27	
2001	584.70	
2002	527.61	511.53
2003	375.19	495.83
2004	524.84	475.88

b)   486.23$/tonne

```
Time       C1    Smooth   Predict      Error
1      422.27   461.061   486.922   -64.652
2      584.70   510.517   461.061   123.639
3      527.61   517.354   510.517    17.093
4      375.19   460.488   517.354  -142.164
5      524.84   486.229   460.488    64.352


Forecasts


Period  Forecast    Lower    Upper
6        486.229  284.402  688.056
```

c)   The absolute percentage error for a) $(577 - 475.88)/577 = 0.1753$ or 17.53%.
The absolute percentage error for b) $(577 - 486.23)/577 = 0.1573$ or 15.73%.

23. Toyota stock prices.

   a)   Irregular (no regular repeating pattern).

   b)   A larger length is used in the second graph (the moving average with the longer length is smoother).

25. Men's weekly earnings. The 2-quarter moving average is shown in A and fits better. This series has a consistent trend component so a longer length moving average is dragged below the series.

27. Autoregressive model.

   a) $\hat{y} = b_0 + b_1 y_{lag1} = 0.100 + 1.12 * 2.3 = \$2.676$ B

   b)   The absolute percentage error (APE) for Q4: $(\$2.90 - \$2.676)/\$2.90 = 0.0772$ or 7.72% (underforecast).

   c) $y_t = 0.410 + 1.35 * 1.9 = \$2.975$ B
   The APE: $\left| (\$2.9 - \$2.975)/\$2.90 \right| = 0.02586$ or 2.6% (overforecast).

29. Coffee prices.

   a)   With a $\hat{y} = b_0 + b_1 y_{lag1} = 1.59175 + 1.22856 * 85.93 - 0.247256 * 82.55 = 86.75$

   b)   The APE: $\left| (86.85 - 86.75)/86.85 \right| = 0.00115$ or 0.115% (underforecast).

   c)   84.24 (from technology): the AR model prediction is closer.

Data	Predict
88.48	
85.31	86.895
78.79	82.05
82.55	80.67
85.93	84.24

31. Gallup poll.

   a)   112 (years since the start of the 20th century: $2012 - 1900 = 112$).

   b) $\widehat{\%RespondingYes} = 5.58 + 0.999 * Year = 5.58 + 0.999 * 112 = 117.47\%$; the forecast will be over 100%, which is not possible.

33. Retail gas, part 2.

  a) There is a linear trend and irregular components.

  b) Prices have increased at about 30 cents per year.

  c) The intercept has no interpretation because year zero has no meaning for this data series.

  d) There is increased variability starting in the middle of 2005.

35. Seasonal model.

  a) $\hat{y} = 1.1 + 0.2t = 1.1 + 0.2*1 = \$1.3\ million$

  b) Q3

  c) Q4

  d) Sales on average are \$0.5 million greater in Q4 than in Q1.

37. Wal-Mart revenue.

  a) A positive trend component and a seasonal component.

  b) Wal-Mart revenues have been increasing at about \$0.145 billion per month.

  c) Revenues in December tend to be about \$11.56 billion more than in January (the base month).

  d) $\overline{Wal-Mart\ Rev\$BPI} = 12.0322 + 0.145241 * 40\ months + 1.46096 = \$19.3028\ billion$

  e) October is the only month in which Wal-Mart revenues are typically lower than they are in January, after allowing for the overall growth trend in revenue.

39. Hawaii tourism.

  a) The P-value indicates that the time trend is significantly different from zero.

  b) September and November (negative coefficients).

  c) July and August (July and August highest coefficient values).

  d) $\widehat{Visitors} = 312,312 + 2016.84 * 63\ months + 20,035.30 = 459,408.22$

41. Oakland passengers.

  a) The log of the number of passengers grew at 0.0249 per month.

  b) In January 1990, there were about $10^{5.64949} = 446,159$ passengers.

  c) January and February; we can tell that January is low because it is the base for the dummy variables and the other coefficients are positive.

  d) Trend, seasonal, and possibly an 8-year cyclic component.

43. Oakland outlier. The single exponential smoothing is pulled down by the outlier. The moving average spreads out the effect of the outlier across several months. The seasonal regression is not noticeably affected (the September dummy coefficient absorbs the effect of the outlier and spreads it out over other Septembers, but that is not evident in the plot).

45. E-commerce, part 2.

a) Dependent variable is: E-Commerce Sales (Millions $)
R squared = 96.9%   R squared (adjusted) = 96.4%
s = 1740 with 34 − 5 = 29 degrees of freedom

Variable	Coeff	SE(Coeff)	t-ratio	P-value
Intercept	2022.8	777.4	2.60	0.014
Time	900.138	30.45	29.6	≤ 0.0001
Q2	−486.223	845.9	−0.575	0.5699
Q3	−709.611	845.4	−0.839	0.4081
Q4	3288.03	820.7	4.01	0.0004

Note: Choosing a different quarter as the base quarter will yield a different model, but one with the same R^2.

b) Dependent variable is: logE-commerce Sales
R squared = 98.7%   R squared (adjusted) = 98.5%
s = 0.0297 with 34 − 5 = 29 degrees of freedom

Variable	Coeff	SE(Coeff)	t-ratio	P-value
Intercept	3.73222	0.0136	274	≤ 0.0001
Time	0.024502	0.0005	47.1	≤ 0.0001
Q2	−2.63785e-3	0.0145	−0.182	0.8566
Q3	−5.43049e-3	0.0145	−0.376	0.7099
Q4	0.072158	0.0140	5.14	≤ 0.000

c) The multiplicative model fits better judging from R^2 or adjusted R^2.

47. Monthly gas prices, part 3.

a) All 4 terms are significant predictors.

```
The regression equation is
Gas = 1.98 + 1.34 Lag  1 - 0.401 Lag  2 + 0.316 Lag  3 - 0.262 Lag  4


278 cases used, 4 cases contain missing values


Predictor      Coef  SE Coef      T     P
Constant      1.977    1.236   1.60  0.111
Lag  1      1.33863  0.05865  22.82  0.000
Lag  2     -0.40111  0.09920  -4.04  0.000
Lag  3      0.31571  0.09951   3.17  0.002
Lag  4     -0.26158  0.05950  -4.40  0.000
```

The regression equation is:
U.S. Retail Gas Price (cents/gal) = 1.98 + 1.34 Lag1 − 0.401 Lag 2 + 0.316 Lag 3 − 0.262 Lag 4.

b) U.S. Retail Gas Price (cents/gal) = 1.98 + 1.34 * 321.1 − 0.401* 306.9 + 0.316 * 300.2 − 0.262 * 292.2. The forecast is $3.27.

c)

Model	Forecast ($)	Actual	APE (%)
Linear	2.38	3.19	25.4
Exponential	2.43	3.19	23.8
AR(4)	3.27	3.19	2.5

d) The AR(4) models has the best fit of those compared here. The APE % is the lowest of all of the models.

49. Oil prices, again.

    a) Linear: Crude Price ($/Bar) = 12.82 + 0.6417*Time = 12.82 + 0.6417*75 = 60.95

    b)

        Exponential: Log (Crude Price ($/Bar) = 1.2437 + 0.00767*Time = 1.2437 + 0.00767*75 = 1.81895

        So Price = $10^{1.81895}$ = 65.94

    c)

Model	Forecast ($)	Actual	APE (%)
Linear	60.95	58.70	3.8
Exponential	65.94	58.70	12.3
AR(1)	53.07	58.70	9.6

    d) The linear fit seems to fit the best because the APE % is the lowest.

51. Mutual funds.

    a) Only the irregular component; this is most likely a random walk (each new value can be thought of as a random step away from the previous value).

    b) Moving Average Models:

c)   Single exponential smoothing model:

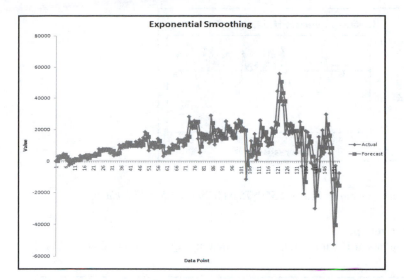

d)

Model	MAPE %	Forecast ($M)	Actual
MA(6)	3.3	–12342.3	–7709
MA(12)	5.3	–27.67	–7709
SES (α = 0.5)	2.0	–15992.8	–7709

e)   MA(6)   MA(12)   SES (α = 0.5)

```
Moving Average          Moving Average          Smoothing Constant

Length  6               Length  12              Alpha  0.5


Accuracy Measures       Accuracy Measures       Accuracy Measures

MAPE          79        MAPE          81        MAPE         103
MAD         6395        MAD         6733        MAD         5838
MSD     116622256       MSD     123443411       MSD     94470753



Forecasts               Forecasts               Forecasts

Period  Forecast        Period  Forecast        Period  Forecast
154     -12342.3        154     -27.6667        154     -15992.8
```

53. Hotel occupancy rates.

a)   Seasonal trend lasting 6 months.

```
Predictor     Coef   SE Coef      T      P
Constant     11.41    10.91    1.05  0.301
Lag 1       0.7641   0.1358    5.63  0.000
Lag 2      -0.2936   0.1737   -1.69  0.098
Lag 3      -0.1030   0.1734   -0.59  0.555
Lag 4       0.3039   0.1733    1.75  0.086
Lag 5      -0.1302   0.1683   -0.77  0.443
Lag 6       0.3049   0.1370    2.23  0.031
```

The only significant predictor values are Lag 1 and Lag 6. The resulting regression:

```
The regression equation is
Hotel = 11.6 + 0.550 Lag 1 + 0.293 Lag 6
```

b)   Residuals over time appear random, except for September 2001.

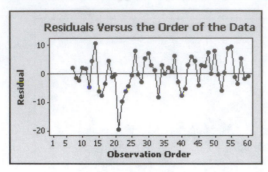

c) $\overline{Hotel\_Occup} = 11.6 + 0.550*75.5 + 0.293*83.5 = 77.59\%$

55. Opec production, part 2.

a)   Analysis of P-values indicate that the only significant predictor is Lag 1.

```
Regression Analysis: Production versus Lag 1, Lag 2, Lag 3, Lag 4

The regression equation is
Production = 786 + 0.920 Lag  1 - 0.102 Lag  2 + 0.047 Lag  3 + 0.110 Lag  4


71 cases used, 4 cases contain missing values


Predictor     Coef  SE Coef      T      P
Constant       786     1026   0.77  0.446
Lag  1      0.9196   0.1205   7.63  0.000
Lag  2     -0.1017   0.1619  -0.63  0.532
Lag  3      0.0472   0.1602   0.29  0.769
Lag  4      0.1099   0.1197   0.92  0.362


S = 557.307   R-Sq = 92.8%   R-Sq(adj) = 92.4%
```

```
Regression Analysis: Production versus Lag 1

The regression equation is
Production = 1145 + 0.963 Lag  1


74 cases used, 1 cases contain missing values


Predictor      Coef  SE Coef      T      P
Constant     1144.8    995.2   1.15  0.254
Lag  1      0.96251  0.03297  29.20  0.000


S = 556.988   R-Sq = 92.2%   R-Sq(adj) = 92.1%
```

The resulting regression equation is: $\overline{OPEC\_Prod} = 1145 + 0.963\text{Lag}1$

b) Forecast for March 2007: $\overline{OPEC\_Prod} = 1144.8 + 0.9625*31{,}286.5 = 31{,}258.06$

c)

Model	Forecast (thousand barrels/day)	Actual	APE (%)
MA(6)	31,689.8	31,286.5	1.29
MA(12)	31,937.5	31,286.5	2.08
SES ($\alpha = 0.9$)	31,299.4	31,286.5	0.04
AR(1)	31,258.06	31,286.5	0.09

d) The single exponential model seems to fit the best because the APE is the lowest.

Chapter 21 –Design and Analysis of Experiments and Observational Studies

SECTION 21.1

  1.

      a) Retrospective

      b) Prospective

SECTION 21.2

  3. The experimental units are chocolate chip cookies. The treatments are the three different kinds of chocolate chips (milk chocolate, dark chocolate and semi-sweet) used in the recipe. The response is ratings (on a scale of 1 to 10) of the chocolate chip cookies by trained taste testers. Randomization was achieved by presenting the chocolate chip cookies to taste testers in a random order.

SECTION 21.3

  5. Control was achieved by keeping all aspects of the recipe (other than the kind of chocolate chips used) the same. Randomization was achieved by presenting the chocolate chip cookies to taste testers in a random order. Replication was achieved by having the same 10 trained testers rate the baked cookies with the different kinds of chocolate chips.

SECTION 21.4

  7.

SECTION 21.5

  9. This experiment is single-blind. Those who could influence the results, in this case the customers at the restaurant were served the new frozen *entre'e* in place of the equivalent *entre'e* without their knowledge. Therefore the customers were blind to the treatments. However, those who would evaluate the results (makers of the new frozen *entre'e*) did know who received the new frozen *entre'e* . Hence, they were not blind to the treatments. Therefore only one group, those who could influence the results, was blind to the treatment.

SECTION 21.6

  11.

      a) For this experiment, there are $k = 4$ treatments and $N = 40$ subjects (4×10). The degrees of freedom for treatment are $k - 1 = 4 - 1 = 3$; degrees of freedom for error are $N - k = 40 - 4 = 36$; and total degrees of freedom are $N - 1 = 40 - 1 = 39$.

      b) $SSE = SSTotal - SST = 1177.97 - 856.07 = 321.9$.

      c) $MST = \dfrac{SST}{k-1} = \dfrac{856.07}{3} = 285.36$.

d) $MSE = \dfrac{SSE}{N-k} = \dfrac{321.9}{36} = 8.94.$

SECTION 21.10

13.

a) There were two factors: *Type of Chocolate Chip* and *Type of Oil* used in the recipe for chocolate chip cookies. The factor levels for the *Type of Chip* were milk chocolate, dark chocolate and semi-sweet and the factor levels for *Type of Oil* were butter or margarine.

b) There are total of 6 (3×2) treatments based on all possible combinations of the factor levels. These are milk chocolate/butter, dark chocolate/butter, semi-sweet/butter, milk chocolate/margarine, dark chocolate/margarine, and semi-sweet/margarine.

c) When an interaction is significant, it means that the effect of one factor on the response variable depends on the level of the other factor. Therefore, we cannot talk about the effects of the factors independently. In this case, for example, it would mean that the relative ratings of the cookies baked with different types of chocolate chips were different for those baked with butter versus margarine.

CHAPTER EXERCISES

15. Laundry detergents.

a) The factors that they are testing: Temperature, Washing Time, and Detergent.

b) The factor levels: Temperature (Hot, Cold), Washing Time (Short, Long), Detergent (Standard, New).

c) The response is the level of whiteness as measured by an optical scanner.

17. Laundry detergents, part 2.

a) Both detergents should be tested under the same conditions to ensure that the results can be compared directly.

b) If cold water washing is eliminated, the results will not be able to be generalized to cold water washing.

c) The treatments should be administered in random order to ensure that no other unforeseen influences could affect the response in a systematic way.

19. Mozart.

a) The difference among the Mozart and the quiet groups was more than would have been expected from ordinary sampling variation.

b)

c) The Mozart group seems shows the smallest median difference and the least improvement. All of the boxes overlap and there does not appear to be a significant difference.

d) No, the difference does not seem significant compared with the usual variation. All of the boxes overlap.

21. Cereal marketing.

    a) Observational; randomly select a group of children, ages 10 to 13, have them taste the cereal and ask them if they like the cereal.

    b) Answers may vary. Get volunteers ages 10 to 13. Each volunteer will be randomly selected to be in one of two groups. Group 1 will taste the new flavor and Group 2 will taste the old flavor. Compare the percentage of favorable ratings for each cereal.

    c) Answers may vary. From the volunteers, identify the children who watch Frump and identify the children who do not watch Frump. Use a blocked design as shown below to focus on the difference. Compare the percentages of favorable ratings for each cereal within the groups of watching Frump or not watching the Frump cartoon.

23. SAT prep courses.

    a) The students were not randomly assigned. Those who signed up for the prep course may be a special group whose scores would have improved anyway.

    b) Answers may vary. Find a group of volunteers willing to participate. Give all subjects the SAT exam. Randomly assign the subjects to the tutorial group or to the no tutorial group. Tutoring is administered to one group. After some time, both groups are retested. Comparison made to see if the tutored group had a significant improvement in scores when compared with the no tutoring group.

    c) After the volunteers have taken the first SAT exam, separate them in blocks of low, average, and high SAT score performance. Randomly assign half of each block to the tutoring and half to the no tutoring groups. Administer the tutoring. Then retest all groups. Compare the differences between treatments for each block.

25. Cars (fuel efficiency).

    a) The null hypothesis H_0: The mean gas Mileage is the same for all cylinder levels ($\mu_4 = \mu_5 = \mu_6 = \mu_8$). The alternative hypothesis H_A: The mean Mileages for all cylinder levels are not all the same.

    b) Randomization not specified, so inference is not clear. The spreads of the four groups look very different. The response variable probably could be re-expressed before proceeding with an Analysis of Variance. Consider omitting the single 5-cylinder car from the study since it is the only one in its group. Boxplots appear symmetric, but other plots should be used to check the Nearly Normal Condition.

27. Cell phone adoption.

    a) The null hypothesis H_0: The mean cell phone adoption rates are equal in the three regions. The alternative hypothesis H_A: The mean cell phone adoption rates are not all equal in the three regions.

b) Countries were selected randomly from the three regions, so the randomization condition is met. However, boxplots suggest the spreads are quite different and show several outliers which require attention. Other plots should be used to check the Nearly Normal Condition.

29. Bank tellers.

a) $H_0: \mu_1 = \mu_2 = \mu_3 = \mu_4 = \mu_5 = \mu_6$ (the mean time it takes to serve a customer is the same for each of the tellers).
H_A: the mean times it takes to serve a customer are not all equal.

b) The P-value of 0.19 is not significant; therefore, we cannot reject the null hypothesis. There is not significant evidence that at least one mean time is different.

c) We cannot perform multiple comparisons because we did not reject the null hypothesis.

31. E-security.

a) This is an observational study. The factor was not deliberately manipulated.

b) This study is retrospective because it uses previously collected data.

c) H_0: The mean responses to the e-security questions are all equal for the three types of community.
H_A: The mean responses to the e-security questions are not all equal for the three types of community.

d) The F-statistic = $\dfrac{MST}{MSE}$ where MST = $\dfrac{SST}{k-1} = \dfrac{6.615}{2} = 3.3075$; MSE = $\dfrac{SSE}{n-k} = \dfrac{96.998}{183} = 0.5300$;

$$\dfrac{MST}{MSE} = \dfrac{3.3075}{0.5300} = 6.24$$

e) The P-value for an F-statistic of 6.24 is 0.002 which is statistically significant. We can reject the null hypothesis and conclude that the mean responses to the e-security question are not all equal for the three types of community. A causal link cannot be established because this is an observational study.

33. Colorful experiment?

a) This is an observational study.

b) There is no treatment. No factors were manipulated.

c) The number of candies of each color.

35. E-trust.

a) This was an experiment. The factors are deliberately manipulated to specify treatments and the subjects are assigned to the treatments randomly.

b) Website trustworthiness was measured on a 10-point scale.

c) There are two factors: 1– website configuration with respect to assurance and 2 – type of product purchased.

d) There are 9 treatments: 3 types of websites and 3 types of purchases.

e) H_0: The mean *trustworthiness* ratings of websites are the same for those with a third-party assurance seal, a self-proclaimed assurance displayed, and no assurance seal.
H_0: The mean *trustworthiness* ratings of websites is the same for online purchases of books, cameras, and insurance.

H_0: There is no interaction effect – the effect of each factor is the same at all levels of the other factor.

37. Stock returns.

    a)   This is an observational study.

    b) H_0: The mean *percentage changes in stock prices* are the same for companies with and without ISO 9000 certification.
        H_0: The mean *percentage changes in stock prices* are the same for large and small companies.
        H_0: There is no interaction effect – the effect of each factor is the same for all levels.

    c)   It is difficult to establish a causal link because this is an observational study. In addition, there is the strong possibility of confounding or lurking variables that are not accounted for in this study. For example, the industry sector may be related to percentage changes in stock prices and ISO 9000 certification but is unaccounted for in this study.

39. Managers' hourly wages.

    a)   This is an observational study. The factors are not deliberately manipulated to specific treatments.

    b)   Boxplots of the hourly wages by both manager type and region indicate that the equal variance assumption is reasonable. Also, the boxplots indicate fairly symmetric distributions. However, other plots should be used to check the Nearly Normal Condition once the model has been fit. The states were randomly selected from the three regions, so the randomization condition is satisfied.

    c)   Based on the P-values, we can conclude that there is no significant interaction effect. There is no evidence that the mean hourly wages are different across the three regions but are different for sales and advertising managers.

    d)   Yes, because the interaction effect is not significant.

41. Production problems.

    a)   This is an experiment. The machine setting factor was deliberately manipulated. Shift is a blocking factor.

    b)   The response variable is the size error of the part.

    c)   There are nine treatments: two factors – three machine settings and three shifts.

    d)   There is no significant interaction effect (P-value is not significant). We can conclude that both the Shift and the Machine Setting appear to have a significant effect (P-values are significant). However, from this table alone, we cannot determine which of these effects is more important.

43. Yogurt research.

    a)   This is an experiment. MST $= \dfrac{SST}{k-1} = \dfrac{17.3}{2} = 8.65$; MSE $= \dfrac{SSE}{n-k} = \dfrac{0.460}{9-3} = 0.0767$.

    b)   The F-statistic $= \dfrac{MST}{MSE} = \dfrac{8.65}{0.0767} = 112.78$

    c)   The data provide very strong evidence that the means are not equal (P-value significant).

    d)   We assumed that the experimental runs were performed in random order, that the variances of the treatment groups are similar, and that the residuals are nearly Normal.

    e) In order to justify the conclusions of the F-test, it would be appropriate to see a boxplot of the *Scores* by *Method*, a plot of residuals vs. predicted values, a Normal probability plot, and a histogram of the residuals.

    f) The average size of the error standard deviation: $s_p = \sqrt{0.0767} = 0.277$ points.

45. Cereal shelf placement.

    a) This is an observational study.

    b) The null hypothesis H_0: The mean *Sugar Content* is the same for the cereals on each *Shelf*. The alternative hypothesis H_A: The means of *Sugar Content* for the cereals are not all equal.

    c) The P-value of 0.0012 provides strong evidence that the mean *Sugar Content* is not the same for each *Shelf*.

    d) We cannot conclude that cereals on *Shelf* 2 have a different mean *Sugar Content* than cereals on *Shelf* 3 or that cereals on *Shelf* 2 have a different mean *Sugar Content* than cereals on *Shelf* 1. We can conclude only that the means are not all equal.

    e) The Bonferroni test shows that at $\alpha = 0.05$ the *Shelves* are different except for *Shelf* 1 and *Shelf* 3, which are not discernibly different. Now we can conclude that the mean *Sugar Content* of cereals on *Shelf* 2 is not equal to the mean *Sugar Content* on *Shelf* 1 and *Shelf* 3. In other words, we can conclude from this test what we wanted to conclude in part d.

47. Automotive safety.

    a) H_0: Head injuries are on the average about the same regardless of the size of the car.
        H_0: Which seat you sit in doesn't affect the degree of head injury.

    b) The conditions appear to be met. We assume the data were collected independently. The boxplots show that the variance is roughly constant, and there are no patterns in the scatterplot of residuals vs. predicted values.

    c) There is no significant interaction. The P-values for both *Seat* and *Size* are < 0.0001. Thus, we reject both of the null hypotheses and conclude that both *Seat* and the *Size* of car affect the severity of head injury.

Chapter 22 – Quality Control

SECTION 22.1

1.

a)

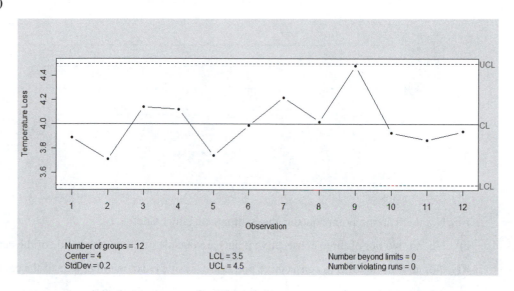

b) The process seems to be in control. There are no points outside the upper or lower specification limits. However, one container (#9) has a temperature loss (4.48 °C) that is very close to the upper specification limit.

c) $C_p = \dfrac{USL - LSL}{6\sigma} = \dfrac{4.5 - 3.5}{6(0.2)} = 0.833$

d) The value of C_p is less than 1.0. This means that the process is not capable of meeting the specification limits. It is usually recommended that the value be 1.25 and sometimes as high as 2.0. Since there is too great a likelihood that some containers will be outside the specification limits for temperature loss, it is recommended that the quality control team make improvements to the process in order to decrease variability in the process (reduce the process standard deviation σ) and increase the value of C_p.

SECTION 22.2

3.

a) 3σ limits: $4 \pm 3(0.2)$: LCL = 3.4 and UCL = 4.6

2σ limits: $4 \pm 2(0.2)$: 3.6 and 4.4

1σ limits: $4 \pm 1(0.2)$: 3.8 and 4.2

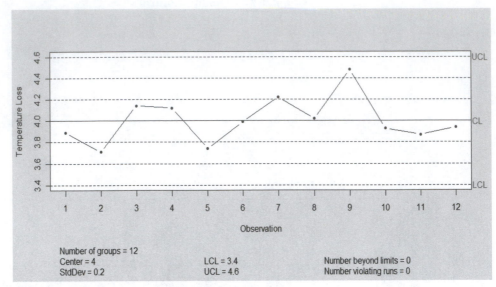

Number of groups = 12
Center = 4
StdDev = 0.2

LCL = 3.4
UCL = 4.6

Number beyond limits = 0
Number violating runs = 0

b) No, there are no points outside the 3σ limits on either side.

c) No, no two out of three consecutive points are outside of 2σ on either side of the centerline.

d) No, no four out of five consecutive points are outside if 1σ on either side of the centerline.

e) No, no eight points are in a row on the same side of the centerline.

f) No, no six points in a row are increasing or decreasing.

g) No, no fourteen points in a row are alternatively increasing and decreasing.

h) Based on the fact that none of the rules for out of control evidence were violated in this control chart (no points indicate that the process is out of statistical control based on the rules defined in parts b – g), we can conclude that the process appears to be stable and in statistical control.

SECTION 22.3

5.

a) The standard deviation of the sample mean is $\dfrac{\sigma}{\sqrt{n}} = \dfrac{0.01m}{\sqrt{5}} = 0.00447m$ or 0.447 cm.

b) The 3σ limits are

$$LCL = \mu - 3\frac{\sigma}{\sqrt{n}} = 91.4 - 3(.00447) = 91.387m$$

$$UCL = \mu + 3\frac{\sigma}{\sqrt{n}} = 91.4 + 3(.00447) = 91.413m$$

Therefore a sample mean would have to be shorter than $91.387m$ or longer than $91.413m$ to fall outside of the 3σ limits.

c) If the sample size is 10, the standard deviation of the sample mean and resulting 3σ limits would change as follows:

$$\frac{\sigma}{\sqrt{n}} = \frac{0.01m}{\sqrt{10}} = 0.00316m$$

$$LCL = \mu - 3\frac{\sigma}{\sqrt{n}} = 91.4 - 3(.00316) = 91.391m$$

$$UCL = \mu + 3\frac{\sigma}{\sqrt{n}} = 91.4 + 3(.00316) = 91.409m$$

Therefore a sample mean would have to be shorter than $91.391m$ or longer than $91.409m$ to fall outside of the 3σ limits. Increasing the sample size decreases the standard deviation of the sample mean and tightens the control limits.

SECTION 22.5

7.

   a) The standard deviation for the sample proportion is $SE(\overline{p}) = \sqrt{\dfrac{pq}{n}} = \sqrt{\dfrac{(.02)(.98)}{50}} = 0.0198$.

   b) The 3σ limits are

$$LCL = \overline{p} - 3\sqrt{\frac{pq}{n}} = 0.02 - 3(.0198) = -0.0394 = 0$$

$$UCL = \overline{p} + 3\sqrt{\frac{pq}{n}} = 0.02 + 3(.0198) = 0.0794$$

   c) The proportion of packages that fails in a box of 50 would have to be above 0.0794 to be declared out of control. This means that 0.0794×50=3.97 or 4 packages out of 50 would have to fail.

CHAPTER EXERCISES

9. **Super Glue®.** These limits are constructed using data collected from a sample of 50 2-oz tubes taken from a process. They are not based on external specifications. Because these limits are used to see how well the process is doing with regard to filling 2-oz tubes, i.e. data collected on the actual weights for a sample of 50 tubes, they are control limits.

11. **Pole vault.** These limits are based on an external specification for the minimum required width and length of pole vault landing pads made by the U.S. National High School Federation (NHSF). They are not based on data collected on the widths and lengths for a sample of pole vault landing pads. Therefore these limits should be treated by manufacturers as lower specification limits for the width and length of landing pads below which no pads should fall.

13. **Observatories.** Gemini has set the requirement for the deviation from flatness of its telescope filters to be no more than 30 *nm* as measured by the root mean square deviation. This requirement is based on the external specification imposed by Gemini. The limits are not based on deviation measures obtained from a sample of filters. Therefore manufacturers who wish to sell their filters for use in telescopes at the two Gemini observatories must view this requirement as an upper specification limit for the deviation from flatness which their filters must not exceed.

15. **Roulette.** This is an example of common-cause variation. The different numbers coming up at the roulette wheel in an evening are due to random fluctuations that are inherent in the process. There is no reason beyond chance for the different numbers that are observed.

17. Out-of-control rules. In order to determine this probability, we need to assume that the variable being charted is normally distributed. Assuming a normal distribution, the probability that an observation will fall beyond 3σ from the mean in a stable process is 0.0026. You can find this using Table Z in the Appendix. The probability that an observation falls outside the 3σ limits is $P(z < -3) + P(z > 3)$. Looking up $z = 3$ in Table Z gives us $P(z \leq 3) = 0.9987$. Therefore, $P(z > 3) = 1 - 0.9987 = 0.0013$. Because of symmetry, $P(z < -3) + P(z > 3) = 2 \times 0.00123 = 0.0026$.

19. Basketballs.

a) Based on the calibration data statistics, the centerline of the control chart is $\overline{\overline{X}} = 8.05$.

   The estimated process standard deviation is $\hat{\sigma} = \dfrac{\overline{R}}{d_2} = \dfrac{0.65}{2.059} = 0.319$.

   The control limits for the $\overline{X}$ chart are

$$LCL = \overline{\overline{X}} - 3\frac{\hat{\sigma}}{\sqrt{n}} = 8.05 - 3 \times \frac{0.319}{\sqrt{4}} = 7.576$$

$$UCL = \overline{\overline{X}} + 3\frac{\hat{\sigma}}{\sqrt{n}} = 8.05 + 3 \times \frac{0.319}{\sqrt{4}} = 8.524$$

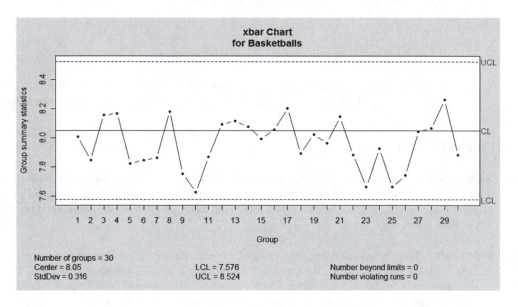

b) Based on the calibration data statistics, the centerline of the control chart is $\overline{R} = 0.65$.

   The control limits for the R chart are

$$LCL = D_3\overline{R} = 0 \times 0.65 = 0$$

$$UCL = D_4\overline{R} = 2.282 \times 0.65 = 1.483$$

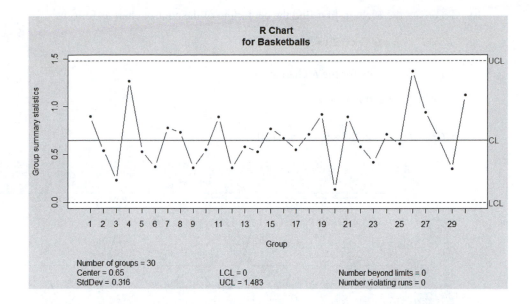

c) Yes, both charts indicate that the company's process for filling its basketballs with the proper amount of air continues to be in control (as defined by the calibration period).

21. Graphite production.

a) Based on specifications for the strength of flexible graphite, $\mu = 21.2$ ounces per square yard with $\sigma = 0.29$ ounces.

The limits for the $\overline{X}$ chart are

$$LCL = \mu - 3\frac{\sigma}{\sqrt{n}} = 21.2 - 3 \times \frac{0.29}{\sqrt{3}} = 20.70$$

$$UCL = \mu + 3\frac{\sigma}{\sqrt{n}} = 21.2 + 3 \times \frac{0.29}{\sqrt{3}} = 21.70$$

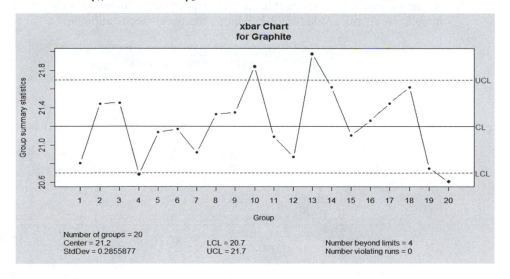

b)   To create the R chart, first find the centerline of the control chart as follows

$$\bar{R} = d_2 \sigma = 1.693 \times 0.29 = 0.491.$$

The control limits for the R chart are

$$LCL = D_3 \bar{R} = 0 \times 0.491 = 0$$

$$UCL = D_4 \bar{R} = 2.282 \times 0.491 = 1.120.$$

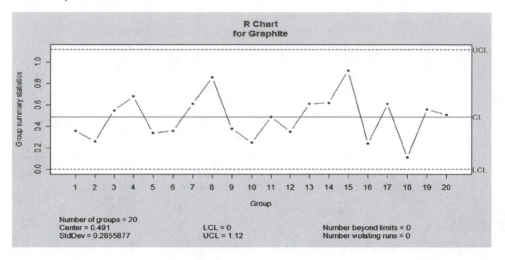

c)   The company's process for making flexible graphite sheets is not in statistical control. While there is no out-of-control evidence in the R chart, the X bar chart shows several samples with mean strengths that are outside the control limits. Samples 2 and 20 are both below the lower control limit with sample mean strengths of 20.69 ounces and 20.61 ounces, respectively. Samples 10 and 13 are both above the upper control limit with sample mean strengths of 21.84 ounces and 21.98 ounces, respectively. Since such occurrences are very unlikely in a stable process, these points should be investigated for assignable special causes.

d)   As can be seen in the runs chart based on the specification limits shown below, all but one of the samples (# 37 with a value of 22.23 ounces) meets the specification limits.

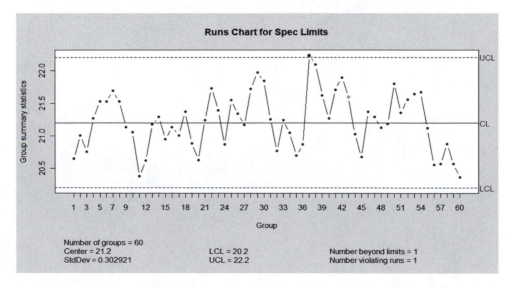

23. Defect monitoring.

a) The centerline of the p chart is

$$\bar{p} = \frac{\hat{p}_1 + \hat{p}_2 + ... + \hat{p}_k}{k} = \frac{\frac{17}{250} + \frac{15}{250} + ... + \frac{18}{250}}{10} = 0.0616$$

b) The upper control limit of the p chart is limits are

$$UCL = \bar{p} + 3\sqrt{\frac{\bar{p}\bar{q}}{n}} = 0.0616 + 3\sqrt{\frac{(0.0616)(1-0.0616)}{250}} = 0.1072.$$

c) The lower control limit of the p chart is.

$$LCL = \bar{p} - 3\sqrt{\frac{\bar{p}\bar{q}}{n}} = 0.0618 - 3\sqrt{\frac{(0.0616)(1-0.0616)}{250}} = 0.0160$$

d)

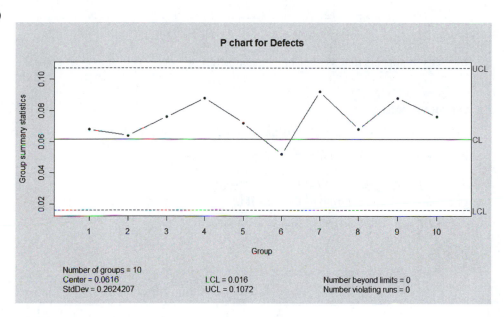

e) The process appears to be in control. No out-of-control evidence is apparent on the control chart.

25. Computer chip manufacturing.

a) The centerline for the c chart is the average number of defects or $\mu = 1.7$ defects per day. The control limits are

$$LCL = \bar{c} - 3\sqrt{\bar{c}} = 1.7 - 3\sqrt{1.7} = -2.21 = 0$$

$$UCL = \bar{c} + 3\sqrt{\bar{c}} = 1.7 + 3\sqrt{1.7} = 5.61$$

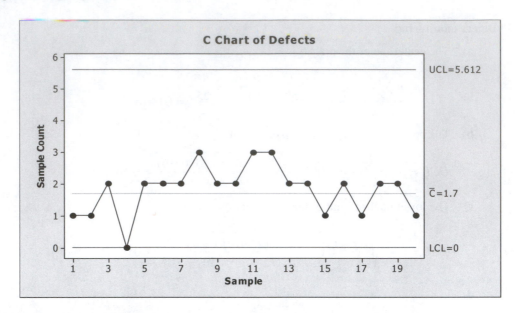

b)   Yes, it appears that MediaChip's manufacturing process is in statistical control.

27. Patient complaints.

a)   The centerline of the c chart is the average number of complaints in the past year or $\mu = 4.57$.

b)   The upper control limit of the c chart is limits is

$$UCL = \bar{c} + 3\sqrt{\bar{c}} = 4.57 + 3\sqrt{4.57} = 10.98$$

c)   The lower control limit of the c chart is

$$LCL = \bar{c} - 3\sqrt{\bar{c}} = 4.57 - 3\sqrt{4.57} = -1.84 = 0$$

d)

e) Yes, this process is out of control. The number of patient complaints on day 5 is above the upper control limit. Since this is an unlikely occurrence in a stable process, management at the small medical facility should examine this day further to identify a potential special cause for this result.

29. Baseball circumferences.

a)

b) The process for making baseballs in terms of their circumference is not in statistical control. Two balls (the first and fifteenth) are below the lower specification limit and therefore too small.

c) There are two balls that are outside the specification limits for circumference. Specifically, balls 1 and 15 are too small.

d) The quality team should have investigated the production process as soon as a ball was found to be outside the specification limits. Since the very first ball selected from this large production batch was below the lower specification limit for circumference, the team should have investigated the process right away.

Chapter 23 – Nonparametric Methods

SECTION 23.1

1.

    a)   Student ratings on a 5-point Likert scale are ranks.   They represent ordered information with no quantitative values or units.

    b)   Students' letter grades in a course are ranks. They represent ordered information with no quantitative values or units.

    c)   Students' point scores on a final exam are not ranks. They represent quantitative data and the units are points.

SECTION 23.2

3.

    a)   No. The Wilcoxon Rank-Sum (or Mann-Whitney) test would <u>not</u> be appropriate in this situation because the *Independent Groups Assumption* is violated. Because the same panel of consumers rates both the new product and its competitor product, the data are paired.

    b)   Yes. The Wilcoxon Rank-Sum (or Mann-Whitney) would be appropriate in this situation. The ratings on sweetness of a diet drink are being compared for two independent groups of tasters (a panel of teenagers and a panel of adults).

    c)   No. The Wilcoxon Rank-Sum (or Mann-Whitney) test would <u>not</u> be appropriate in this situation because it does not involve a comparison of two independent groups. It involves determining the presence of a consistent trend (or monotone) of water heights with changing moon phases.

    d)   No. The Wilcoxon Rank-Sum (or Mann-Whitney) test would <u>not</u> be appropriate in this situation because there are more than two groups (there are four groups) being compared.

    e)   No. The Wilcoxon Rank-Sum (or Mann-Whitney) test would <u>not</u> be appropriate in this situation because it involves computing a measure of association between two variables (home value and age) as opposed to comparing two independent groups.

SECTION 23.3

5.

    a)   No. The Kruskal-Wallis test would <u>not</u> be appropriate in this situation because there are only two groups. Furthermore, the *Independent Groups Assumption* is violated because the same panel of consumers rates both the new product and its competitor product.

    b)   No. The Kruskal-Wallis test would <u>not</u> be appropriate in this situation because the ratings on sweetness of a diet drink are being compared for only two independent groups of tasters (a panel of teenagers and a panel of adults).

    c)   No. The Kruskal-Wallis test would <u>not</u> be appropriate in this situation because it does not involve a comparison of more than two independent groups. It involves determining the presence of a consistent trend (or monotone) of water heights with changing moon phases.

    d)   Yes. The Kruskal-Wallis test would be appropriate in this situation because the ratings of a new distance learning course are being compared among four independent groups (Freshmen, Sophomores, Junior and Seniors).

    e)   No. The Kruskal-Wallis test would <u>not</u> be appropriate in this situation because it involves computing a measure of association between two variables (home value and age) as opposed to comparing more than two independent groups.

SECTION 23.4

7.

a) Yes. The Wilcoxon Signed-Rank test would be appropriate in this situation because two groups are being compared and the data are naturally paired (because the same panel of consumers rates both the new product and its competitor product).

b) No. The Wilcoxon Signed-Rank test would <u>not</u> be appropriate in this situation because the ratings on sweetness of a diet drink are being compared for two independent groups of tasters (a panel of teenagers and a panel of adults). The data are not appropriately paired.

c) No. The Wilcoxon Signed-Rank test would <u>not</u> be appropriate in this situation because it does not involve a comparison of two groups for which the data are paired. It involves determining the presence of a consistent trend (or monotone) of water heights with changing moon phases.

d) No. The Wilcoxon Signed-Rank test would <u>not</u> be appropriate in this situation because the ratings of a new distance learning course are being compared among four groups (Freshmen, Sophomores, Junior and Seniors) that are independent.

e) No. The Wilcoxon Signed-Rank test would <u>not</u> be appropriate in this situation because it involves computing a measure of association between two variables (home value and age) as opposed to comparing two groups for which the data are paired.

SECTION 23.6

9.

a) No. Kendall's tau would <u>not</u> be appropriate in this situation because the interest is in comparing two groups with paired data (the same panel of consumers rates both the new product and its competitor product) and not in determining whether there is a monotone association between two variables.

b) No. Kendall's tau would <u>not</u> be appropriate in this situation because the interest is in comparing the ratings on sweetness of a diet drink for two independent groups of tasters (a panel of teenagers and a panel of adults) and not in determining whether there is a monotone association between two variables.

c) Yes. Kendall's tau would be appropriate in this situation because it involves determining the presence of a consistent trend (or monotone) of water heights with changing moon phases.

d) No. Kendall's tau would <u>not</u> be appropriate in this situation because the interest is in comparing ratings of a new distance learning course among four independent groups (Freshmen, Sophomores, Junior and Seniors) and not in determining whether there is a monotone association between two variables.

e) Maybe. Kendall's tau could be appropriate in this situation if the interest was in determining whether the association between the two variables (home value and age) is monotone. However, a better measure of association in this case is probably Spearman's rho.

SECTION 23.7

11.

a) No. Spearman's rho would <u>not</u> be appropriate in this situation because the interest is in comparing two groups with paired data (the same panel of consumers rates both the new product and its competitor product) and not in computing a measure of association between two variables.

b) No. Spearman's rho would <u>not</u> be appropriate in this situation because the interest is in comparing the ratings on sweetness of a diet drink for two independent groups of tasters (a panel of teenagers and a panel of adults) and not in computing a measure of association between two variables.

c) Maybe. Spearman's rho could be appropriate in this situation if the data could be ranked. However, Kendall's tau would be more appropriate in this situation because the interest is in determining the presence of a consistent trend (or monotone) of water heights with changing moon phases.

d) No. Spearman's rho would <u>not</u> be appropriate in this situation because the interest is in comparing ratings of a new distance learning course among four independent groups (Freshmen, Sophomores, Junior and Seniors) and not in computing a measure of association between two variables.

e) Yes. Spearman's rho would be appropriate in this situation since the interest is in measuring the association between the two variables (home value and age).

CHAPTER EXERCISES

13. Product testing.

a) The assumptions and conditions are met for the Wilcoxon Rank Sum (Mann-Whitney) test as the groups are independent and children are randomly assigned to the groups. We can state the hypotheses as

H_0: The new reading activities and the traditional teaching methods yield the same typical results on reading comprehension for elementary school children.

H_A: The new reading activities yield better typical results on reading comprehension for elementary school children compared to traditional teaching methods.

Results of the Mann-Whitney test using technology:

```
Mann-Whitney Test and CI: New Activities, Control
                   N   Median
New Activities    18    53.50
Control           22    42.00

W = 462.0
Test of ETA1 = ETA2 vs ETA1 > ETA2 is significant at 0.0060
The test is significant at 0.0059 (adjusted for ties)
```
Based on these results, we can reject the null hypothesis and conclude that the new reading activities improve reading comprehension for elementary school children.

b) It would be better to design this as a before/after study, testing the children at the start of the eight week program, randomly assigning them to the treatment groups (Group 1: new reading activities are used; Group 2: traditional teaching methods), and then testing them at the end of the eight week program. This type of design controls for individual differences in reading comprehension among students. An appropriate nonparametric for comparing results from this type of study would be the Wilcoxon Signed-Rank test.

15. Cereal sugar.

a) The null and alternative hypotheses are

H_0: Children's cereals and adult cereals have the same typical sugar content (the distributions of sugar content have the same center).

H_A: Children's cereals and adult cereals differ in typical sugar content (the distributions of sugar content have different centers).

b) The independent groups assumption required for the Wilcoxon Rank Sum (Mann-Whitney) test is satisfied. We don't know if these cereals are a random sample from any particular population, but they are likely to be mutually independent.

c)   Results of the Mann-Whitney test using technology:

```
Mann-Whitney Test and CI: Children, Adult
            N   Median
Children   19   45.900
Adult      28    7.950

W = 722.0
Test of ETA1 = ETA2 vs ETA1 not = ETA2 is significant at 0.0000
The test is significant at 0.0000 (adjusted for ties)
```

Based on these results, we can reject the null hypothesis and conclude that there is a difference in sugar contents between children and adult cereals.

d)

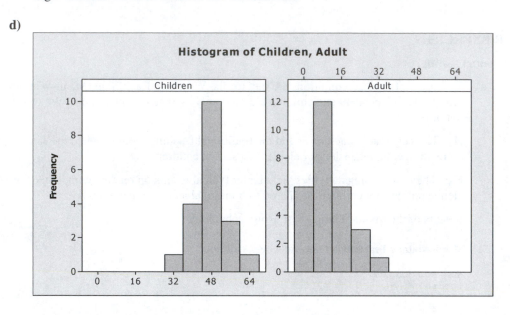

The distribution of sugar content in adult cereals is skewed to the right. The distribution of sugar content in children's cereals is symmetric. The skewness is a mild violation of the nearly Normal condition required for the two-sample t test. We can re-express the sugar content of adult cereals in order to make the distribution symmetric. However, we cannot re-express one set of data without re-expressing the other, and such a re-expression would make the symmetric distribution of children's cereals sugar content skewed. Therefore, the nonparametric Wilcoxon Rank Sum (Mann-Whitney) test is a better choice than the two sample t test for these data.

17. Trophy sizes. The two groups (size of trophies created in the 19th century and size of trophies created in the 20th century) are independent so the appropriate nonparametric method is the Wilcoxon Rank Sum (Mann-Whitney) test. We can state the hypotheses as

H_0: Trophies created in the 19th and 20th centuries have the same typical size (the distributions of trophy sizes have the same center).

H_A: Trophies created in the 19th and 20th centuries differ in typical size (the distributions of trophy sizes have different centers).

Results of the Mann-Whitney test using technology:

```
Mann-Whitney Test and CI: Size19th, Size20th
            N   Median
Size19th    5   27.00
Size20th    9   15.50

W = 54.5
Test of ETA1 = ETA2 vs ETA1 not = ETA2 is significant at 0.0278
The test is significant at 0.0273 (adjusted for ties)
```

Based on these results, we can reject the null hypothesis. The P-value of 0.0278 is less than $\alpha = 0.10$. We can conclude that the typical size of trophies created in the 19th and 20th centuries was not the same.

19. Freshman 15.

a) The data are appropriately paired as each student is weighed the first week of the semester and then again at the end of 12 weeks. While the students are not randomly selected (volunteers), they do match the rest of the freshman class in terms of demographic variables and are therefore representative. They are almost surely independent of each other. The conditions and assumptions for the Wilcoxon Signed-Rank test appear reasonably satisfied. The hypotheses are

H_0: There is no change in students' weights over time (the median change is zero).

H_A: There is a change in students' weights over time (the median change is not zero).

Results of the Wilcoxon Signed-Rank test using technology:

Wilcoxon Signed Rank Test: Diff(Term-Init)
```
Test of median = 0.000000 versus median not = 0.000000

                     N for   Wilcoxon             Estimated
                N    Test    Statistic     P       Median
Diff(Term-Init) 68    57      1526.5     0.000     2.000
```

Based on the results, we can reject the null hypothesis and conclude that the students' weights did change over time. The estimated median change in the difference is 2 pounds (note that the difference was calculated as *Terminal Weight – Initial Weight*). Consequently, the change in students' weights is in the expected direction.

b) It appears that the assumptions and conditions for the paired *t* test are reasonably satisfied by these data. Below is a histogram of the differences which shows that they are nearly Normal (aside from an apparent outlier on the low end). The paired *t* test may be a better choice for these data as it also would provide a confidence interval for the weight change.

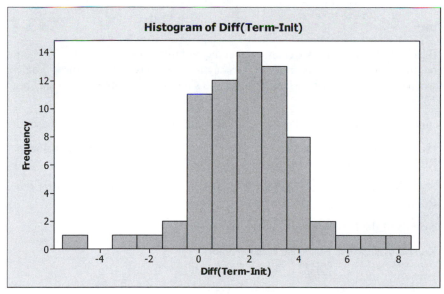

21. ERP effectiveness. The data are appropriately paired (times for financial close are measured for the same set of companies before and after implementing ERP). We don't know if this sample of companies is random but it is reasonable to assume that they are independent of each other. Using the Wilcoxon Signed Rank test we state the hypotheses as

H_0: There is no change in the time it takes to complete the financial close process for companies as a result of implementing ERP (the median change is zero).

H_A: There is acceleration in the time it takes to complete the financial close process for companies as a result of implementing ERP (the median change, *After – Before*, is negative).

Results of the Wilcoxon Signed-Rank test using technology:

Wilcoxon Signed Rank Test: Diff(After-Before)

```
Test of median = 0.000000 versus median < 0.000000

                      N for   Wilcoxon            Estimated
                   N  Test  Statistic      P       Median
Diff(After-Before) 8     7        2.5  0.031      -0.9500
```

Based on these results, we can reject the null hypothesis. The P-value of 0.031 is less than typical α values (e.g., 0.05). We conclude that implementing ERP appears to be effective in accelerating a company's financial close process.

23. **Friday the 13th, traffic.** The data are appropriately paired (cars are counted on two different Fridays of the same month). The months are not from one season, so we may consider them representative of different times over the year and relatively independent of each other. The Wilcoxon Signed Rank test is appropriate. The relevant hypotheses are

H_0: There is no difference in the number of cars passing the junctions on the motorway on consecutive Fridays of the month when one is the 13th (the median difference is zero).

H_A: There is a difference in the number of cars passing the junctions on the motorway on consecutive Fridays of the month when one is the 13th (the median difference is not zero).

Results of the Wilcoxon Signed-Rank test using technology:

Wilcoxon Signed Rank Test: Diff

```
Test of median = 0.000000 versus median not = 0.000000
            N for   Wilcoxon            Estimated
         N  Test  Statistic      P       Median
Diff  5     5       15.0  0.059         1889
```

Based on these results, we cannot reject the null hypothesis at α=0.05. We conclude that there is no difference in the number of cars passing the junctions on the motorway on consecutive Fridays when one is the 13th (there is no evidence to suggest that people tend to stay home on Friday the 13th).

25. **Cuckoos and quality control.** Using the Kruskal-Wallis test, the relevant hypotheses are

H_0: There is no difference in the size of cuckoo eggs among the three foster species (Sparrow, Robin or Wagtail).

H_A: There is a difference in the size of cuckoo eggs among the three foster species (Sparrow, Robin or Wagtail).

Results of the Kruskal-Wallis test using technology:

Kruskal-Wallis Test: Response versus Factor

```
Kruskal-Wallis Test on Response
Factor    N  Median  Ave Rank       Z
Robin    15   22.45      18.3   -1.26
Sparrow  13   23.05      24.7    1.13
Wagtail  14   22.85      22.0    0.17
Overall  42              21.5

H = 1.92  DF = 2  P = 0.383
H = 1.96  DF = 2  P = 0.376  (adjusted for ties)
```

Based on these results, we cannot reject the null hypothesis at $\alpha=0.05$. We conclude that there is no difference in the size of cuckoo eggs among the three different foster species (there is no evidence to suggest that the cuckoo changes the size of her eggs for different foster species).

27.   Crowdedness.

    a)   The value of -0.626 for Kendall's tau tells us that the relationship between *Crowdedness* (average number of persons per room in a home) and a country's *GDP* is negative and strongly monotonic.

    b)   The value of -0.804 for Spearman's rho tells us that there is a strong negative association between *Crowdedness* and a country's *GDP*.

    c)   From the scatterplot it is apparent that the relationship between *Crowdedness* and a country's *GDP* is not linear. The Pearson correlation is very sensitive to violations of the Linearity Condition. Therefore it is more appropriate to use nonparametric measures to describe this relationship.

29.   The Home Depot.

    a)   The scatterplot of quarterly *Sales* at The Home Depot versus quarterly *Housing Starts* shows that the Linearity condition is not satisfied as there appears to be some curvature in the plot (a bend in the data). Because the Pearson correlation is very sensitive to violations of the Linearity condition, Kendall's tau is preferable.

    b)   Since Kendall's tau records whether the slope between pairs of points is positive, negative or zero it tells us how consistently sales at the The Home Depot grow as housing starts grow.

31. Income and housing. The value of 0.51 for Kendall's tau tells us that there is a fairly consistent (monotonic) positive association between the *Housing Cost Index* and the *Median Family Income* at the state level. The correlation of 0.65 indicates a moderate positive linear association between the two variables. However, neither necessarily implies a cause and effect relationship. Therefore it is not appropriate to conclude that raising the median family income in a state will raise its *Housing Cost Index*.

Chapter 24 –Decision Making and Risk

SECTION 24.1

1.

    a) Whether employment is above or below 10% is a state of nature. This is a fact about the world that affects the consequences of actions.

    b) Whether a new product should be brought to market before or after the beginning of the fiscal year is an action. These are alternative choices of a decision.

    c) Whether to tell employees to stay home because of a blizzard is an action. These are alternative choices of a decision.

SECTION 24.2

3.

The decision tree for deciding how your company should produce its new cell phone screen defroster:

Action	State	Outcome
Out-source	Recession	220
	Stable	350
	Expansion	300
In-House	Recession	150
	Stable	240
	Expansion	390

SECTION 24.3

5.

    a) The minimax choice is *Out-source*. This choice minimizes the maximum cost since the maximum cost for *Out-source* ($300,000) is less than the maximum cost for *In-house* ($390,000).

    b) The minimin choice is *In-house*. This choice minimizes the minimum cost since the minimum cost for *In-house* ($150,000) is less than the minimum cost for *Out-source* ($220,000).

SECTION 24.4

7.

    a) $EV(\text{Out-source}) = 0.2 \times 220 + 0.2 \times 350 + 0.6 \times 300 = 294$ or $294,000
$EV(\text{In-house}) = 0.2 \times 150 + 0.2 \times 240 + 0.6 \times 390 = 312$ or $312,000

    b) Since we wish to minimize expected cost, the best choice using the expected value approach is to outsource production of the new cell phone screen defroster.

SECTION 24.7

9.

$SD(\text{Out-source}) =$

$$\sqrt{(220-294)^2 \times 0.2 + (350-294)^2 \times 0.2 + (300-294)^2 \times 0.6} = \sqrt{1744} = 41.761 \text{ or } \$41,761$$

$$CV(Out-source) = \frac{SD(Out-source)}{EV(Out-source)} = \frac{41.761}{294} = 0.14$$

$SD(\text{In-house}) =$

$$\sqrt{(150-312)^2 \times 0.2 + (240-312)^2 \times 0.2 + (390-312)^2 \times 0.6} = \sqrt{9936} = 99.679 \text{ or } \$99,679$$

$$CV(In-house) = \frac{SD(In-house)}{EV(In-house)} = \frac{99.679}{312} = 0.32$$

SECTION 24.10

11.

a)

b)   P(*Hispanic/Latino* and *Female*) = P(*Hispanic/Latino*)×P(*Female* | *Hispanic/Latino*)
P(*Hispanic/Latino* and *Female*) = 0.13 × 0.42 = 0.0546

c)   P(*Female* | *White*) = 0.46

SECTION 24.11

13.

$$P(White / Female) = \frac{P(Female / White) \times P(White)}{P(Female / White) \times P(White) + P(Female / Black) \times P(Black) + P(Female / Hispanic) \times P(Hispanic)}$$

$$P(White / Female = \frac{0.46 \times 0.68}{0.46 \times 0.68 + 0.48 \times 0.11 + 0.42 \times 0.13} = \frac{0.3128}{0.4202} = 0.744$$

CHAPTER EXERCISES

15. **Flight decision.**

		Stay for Interview	
		No	**Yes**
Action	**Fully Changeable Fare**	$750	$750
	Non-refundable Fare	$650	$800

17. **Advertising strategies.**

19. Flight decision tree.

21. Flight decision expected value.

EV (Fully Changeable Fare) = 0.30 * $750 + 0.70 * $750 = $750.
EV (Non-refundable Fare) = 0.30 * $800 + 0.70 * $650 = $695. Better to choose the non-refundable fare in order to minimize the expected cost.

23. Flight decision, part 4.

EV (Fully Changeable Fare) = 0.70 * $750 + 0.30 * $750 = $750.
EV (Not changeable) is 0.70 * $800 + 0.30 * $650 = $755, so the changeable fare is a slightly better choice.

25. Advertising strategies decisions.

    a) If P (rising confidence) = 0.70, EV (Prime Time) = 0.70*$20 + 0.30*$2 = $14.60.
    E (Targeted Web Marketing) = 0.70*$12 + 0.30*$10 = $11.40.
    EV (Direct Mail) = 0.70*$10 + 0.30*$15 = $11.50.
    Choose Prime Time since expected payoff is highest.

    b) If P (rising confidence) = 0.40, EV (Prime Time) = 0.40*$20 + 0.60*$2 = $9.20.
    E (Targeted Web Marketing) = 0.40*$12 + 0.60*$10 = $10.80.
    EV (Direct Mail) = 0.40*$10 + 0.60*$15 = $13.00.
    Choose Direct Mail in this case since expected payoff is highest.

27. Advertising strategies EVPI.

    a) If P (rising confidence) = 0.70,
    Prime Time Ad: 0.70*$20 + 0.30*$2 = $14.60
    Targeted Web Marketing: 0.70*$12 + 0.30*$10 = $11.40
    Direct Mail: 0.70*$10 + 0.30*$15 = $11.50
    $EVwPI = 0.7*\$20 + 0.30*\$15 = \$18.50$
    $EVPI = |EVwPI - EV| = |\$18.50 - \$14.60| = \$3.90$

    b) If P (rising confidence) = 0.40,
    EV (Prime Time) = 0.40*$20 + 0.60*$2 = $9.20
    E (Targeted Web Marketing) = 0.40*$12 + 0.60*$10 = $10.80
    EV (Direct Mail) = 0.40*$10 + 0.60*$15 = $13.00
    $EVwPI = 0.4*\$20 + 0.60*\$15 = \$17.00$
    $EVPI = |EVwPI - EV| = |\$17.00 - \$13.00| = \$4.00$

29. Advertising strategies with information.

a)

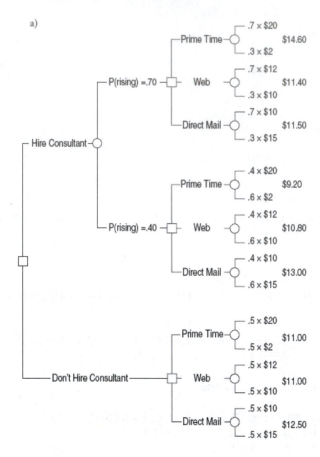

b)   Yes, both alternatives with information have a better return.

c)   Compare 50-50: 0.5*$11.40 + 0.5*$9.20 = $10.30; 70-40: 0.7*$11.40 + 0.4*$9.20 = $11.66: Difference = $11.66 – $10.30 = $1.36.

31. Investing in equipment.

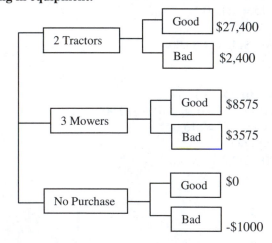

	Growing Season	
Purchase Decision	Good	Bad
2 Tractors	$27,400	$2,400
3 Mowers	$8575	$3575
No Purchase	$0	−$1000

33. Investing in equipment, maxes and mins. Shawn favors Tractors, with a positive gain of $27,400. Lance prefers Mowers, with the worst result being $3575.

35. "Weather" or not to invest in equipment.

<table>
<thead>
<tr><th></th><th></th><th>EV</th><th>SD</th><th>RRR</th></tr>
</thead>
<tbody>
<tr><td rowspan="3">Purchase Decision</td><td>2 Tractors</td><td>$19,900.00</td><td>11,456.44</td><td>1.737</td></tr>
<tr><td>3 Mowers</td><td>$7075</td><td>2291.28</td><td>3.088</td></tr>
<tr><td>No Purchase</td><td>−$300.00</td><td>458.26</td><td>−0.655</td></tr>
</tbody>
</table>

a) EV (2 Tractors) = 0.7*$27,400 + 0.3*$2,400 = $19,900;
EV (3 Mowers) = 0.7*$8,575 + 0.3*$3575 = $7075;
EV (No Purchase) = 0.7*$0 + 0.3*(-$1000) = -$300

b) SD (2 Tractors) = $\sqrt{(\$27,400-\$19,900)^2*0.7+(\$2,400-\$19,900)^2*0.3}$ = $11,456.44

SD (3 Mowers) = $\sqrt{(\$8,575-\$7075)^2*0.7+(\$3575-\$7075)^2*0.3}$ = $2,291.28

SD (No Purchase) = $\sqrt{(\$0-(-\$300))^2*0.7+((-\$1,000)-(-\$300))^2*0.3}$ = $458.26

c) RRR (2 Tractors) = EV/SD = $19,900/$11,456.44 = 1.737;
RRR (3 Mowers) = EV/SD = $7,075/$2,291.28 = 3.088;
RRR (No Purchase) = EV/SD = (-$300.00)/$458.26 = -0.655;
Purchasing Mowers has the highest RRR, which is preferred.

37. Equipment and data.

23. a)

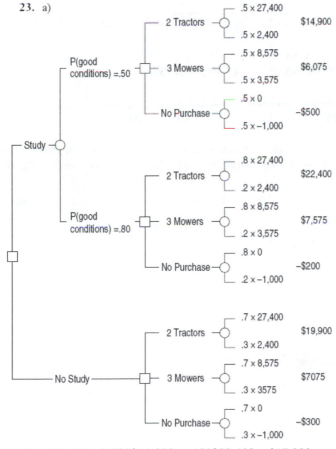

b) EVwSI = 0.60*$14,900 + .40*$22,400 = $17,900

c) No. In all scenarios, the Tractors investments seems the best course to take.

39. Investment strategy.

a) EV (Stock A) = 0.3*$5000 + 0.7*-$1000 = $800;
EV (Stock B) = 0.3*(-$500) + 0.7*$3500 = $2300

b) SD (Stock A) = $\sqrt{(\$5000-\$800)^\wedge 2*0.3+(-\$1000-\$800)^\wedge 2*0.7}=\$2749.55$

SD (Stock B) = $\sqrt{(-\$500-\$2300)^\wedge 2*0.3+(\$3500-\$2300)^\wedge 2*0.7}=\$1833.03$

c) CV (Stock A) = SD/EV = $2749.55/$800 = 3.437;
CV (Stock B) = SD/EV = $1833.03/$2300 = 0.797

RRR (Stock A) = EV/SD = $800/$2749.55 = 0.291;
RRR (Stock B) = EV/SD = $2300/$1855.03 = 1.255

d) Stock B; it has a higher expected value and a lower SD. Its RRR is much higher which shows it has a better reward to risk ratio.

41. Bicycle sales.

a) EV (High-End Bike) = 0.3*$1900 + 0.5*$4750 + 0.2*$7600 = $4465;
EV (Moderately-Priced Bike) = 0.3*$4000 + 0.5*$2500 + 0.2*$1000 = $2650

b) SD (High-End Bike) =
$\sqrt{(\$1900-\$4465)^\wedge 2*0.3+(\$4750-\$4465)^\wedge 2*0.5+(\$7600-\$4465)^\wedge 2*0.2}=\$1995$;
SD (Moderately-Priced Bike) =
$\sqrt{(\$4000-\$2650)^\wedge 2*0.3+(\$2500-\$2650)^\wedge 2*0.5+(\$1000-\$2650)^\wedge 2*0.2}=\$1050$

c) CV (High-End Bike) = SD/EV = $1995/$4465 = 0.447;
CV (Moderately-Priced Bike) = SD/EV = $1050/$2650 = 0.396

RRR (High-End Bike) = EV/SD = $4465/$1995 = 2.238;
RRR (Moderately-Priced Bike) = EV/SD = $2650/$1050 = 2.524

d) Answers may vary. The high-end bike has a much higher EV but also has a higher SD. Its RRR is lower but is very close to the moderately-priced bike. The decision will depend on the store owner's acceptance of risk.

43. Website experiment.

a)

b) The percent of customers who visited the site and made a purchase: $0.012 + 0.015 + 0.01 = 0.037$ or 3.7%.

c) The probability the customers were offered free shipping given that they made a purchase:
$$\frac{0.015}{0.012+0.015+0.01} = 0.405$$

45. Contract bidding.

a)

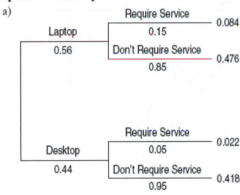

b) Probability you will get both contracts is $0.8*0.2 = 0.16$.

c) Probability that you got the first contract, given that you got the second contract:
$$\frac{0.16}{0.16+0.08} = 0.667$$

47. Computer reliability.

a)

Laptop 0.56
  Require Service 0.15 — 0.084
  Don't Require Service 0.85 — 0.476

Desktop 0.44
  Require Service 0.05 — 0.022
  Don't Require Service 0.95 — 0.418

b) The percentage of computers sold by the bookstore last semester requiring service: $0.084 + 0.022 = 0.106 = 10.6\%$.

c) Percentage of customers owning laptops given that their computer required service:
$$\frac{0.084}{0.084+0.022} = 0.792 = 79.2\%.$$